PLANNING FOR WIND ENERGY

Suzanne Rynne, AICP, Larry Flowers, Eric Lantz,
and Erica Heller, AICP, editors

TABLE OF CONTENTS

CHAPTER 1

Introduction

Larry Flowers with Eric Lantz

 For centuries, the power of the wind has been harnessed for the benefit of humanity and commerce. In the United States, mechanical wind systems pumped water and helped open the Great Plains to human settlement and agricultural production during the 1800s. In the early 20th century, "wind chargers" brought lights and communication technology to rural American households and businesses. Wind energy provided electricity to rural markets prior to the development of federal hydropower dams, the associated interstate transmission system, and the Rural Electrification Administration.

In the early 1980s, wind energy first began to penetrate wholesale electricity markets. In California, Governor Jerry Brown implemented incentives that, in combination with federal tax credits and the federal Public Utilities Regulatory Policy Act (PURPA), launched the contemporary wind electricity era. While the turbines were initially small (on the order of 50 kilowatts) and less reliable than they are today, they were the genesis of a dynamic and robust industry. Improvements in technology resulted in larger, more efficient, and more reliable designs, as well as increased use (Figure 1.1). Today, the cost of wind energy is approximately 25 to 35 percent of what it was in the early 1980s (DEA 1999, Krohn 2009).

Figure 1.1. *The growth of wind turbine size and capacity, 1980–2010*

Source: NREL

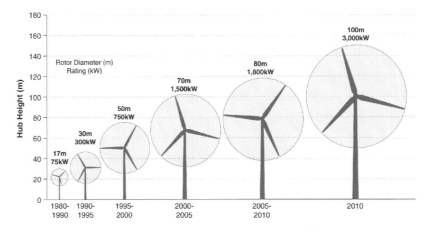

In the first decade of the 21st century, wind energy moved into the mainstream, emerging as a significant source of power generation in Europe, the United States, and Asia (Figure 1.2). In the United States between 2000 and 2010, wind energy generation capacity grew from 2,500 megawatts (MW) to 40,000 MW (Wiser and Bolinger 2011). The number of U.S. states with at least 100 MW of installed capacity grew from four to 28, and half of those now have more than 1,000 MW installed capacity (Figure 1.3; Wiser and Bolinger 2011). Worldwide, installed capacity has grown to more than 194,000 MW; China and India together account for more than 55,000 MW (GWEC 2011).

Figure 1.2. *Annual and cumulative installed wind power capacity in the United States*

Source: Wiser and Bolinger 2011

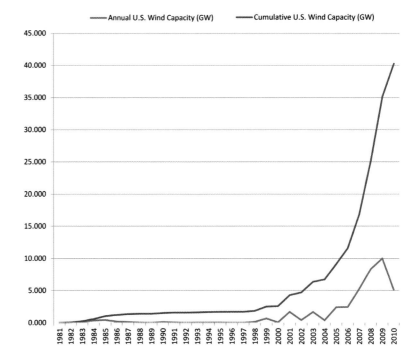

Over the past decade, wind energy has become increasingly competitive with other sources of electricity. With natural gas prices increasing throughout much of the early 2000s, wind became directly competitive in many regions. Regulatory commissions, which need to balance rate increases, electricity reliability requirements, and utility financial returns, have been increasingly compelled to consider wind as an option in utility generation plans. Although falling natural gas prices and the emergence of wholesale electricity markets have made it more challenging for wind to compete in the past, continued technology and improvements and production efficiencies suggest that wind is likely to maintain its competitive position over the long term (Wiser and Bolinger 2011).

THE BENEFITS OF WIND

The environmental benefits of wind energy are numerous and significant—and for many advocates and green power purchasers they are the primary motivation for supporting wind energy. Wind energy, unlike fossil fuels, produces no sulfur monoxide, nitric oxide, particulates, carbon dioxide, or mercury. These fossil-fuel pollutants contribute to acid rain, smog, asthma, climate change, and water pollution. Furthermore, wind energy requires negligible amounts of water, whereas thermal generators (including nuclear) are among the largest consumers of water in the United States. Water is a finite resource, and in many parts of the country water is becoming a critical limiting resource. This is especially true in the American West, where population growth and climate change are expected to place increasing pressure on scarce water resources. Additionally, wind energy requires no mining, drilling, or transportation of fuel, it creates no hazardous waste, and it poses no risk of large-scale environmental contamination.

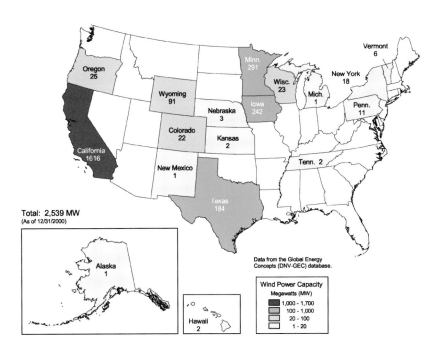

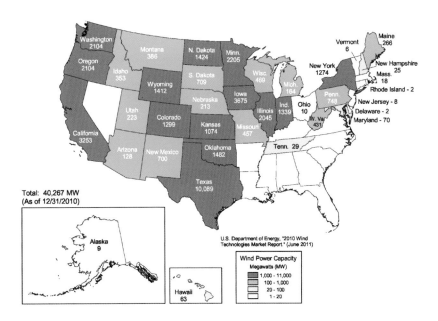

Figure 1.3. Installed U.S. wind power capacity (MW) by state in 2000 and 2010

Source: NREL

A number of global events have made energy security an increasingly critical topic, including the oil price shocks of the 1970s and, more recently, increasing awareness of the West's significant dependence on foreign sources of oil. Recent and ongoing dramatic economic growth in China and India has placed further demands on the international fossil-fuel supply. As a result, homegrown resources, especially those that are infinitely renewable, are increasingly appealing to government officials, business leaders, and consumers. Moreover, domestic wind energy potential could theoretically provide more than nine times the nation's current electricity use (Elliot et al. 2010; EIA 2010). Only a handful of states do not have developable wind resources (whether land-based or offshore), and all states have access to wind energy through the interstate transmission system.

The economic development benefits of wind energy may be the most tangible basis for local and state officials' interests in it. In addition to the direct salaries associated with building and operating wind projects, the wind energy industry provides indirect jobs and benefits (e.g., component and material suppliers, financing and banking, landowner lease payments, and property taxes) and induced jobs (e.g., in local shops, transit, day care, and medical facilities). For example, the first 1,000 MW of wind developed in Colorado produced 1,700 construction-related job-years and 300 permanent jobs, and the total impact on local economies over 20 years is expected to be $975 million (Reategui and Tegen 2008). At a time when America is economically stressed, the new investment and jobs brought by wind energy projects are highly valued by state and local officials and businesses.

When a state lands a manufacturer of a major wind component (e.g., blades, towers, nacelles) or converts an existing manufacturing facility to make or process subcomponents, it further benefits from wind deployment in the form of skilled manufacturing jobs. As an example, Iowa—the state with the second largest amount of installed wind generating capacity—has attracted a number of wind manufacturing facilities. Figure 1.4 illustrates

Figure 1.4. Economic development impacts from increased utilization of Iowa-based wind turbine manufacturing (assuming 2,400 MW of deployments)

Source: Lantz and Tegen 2008

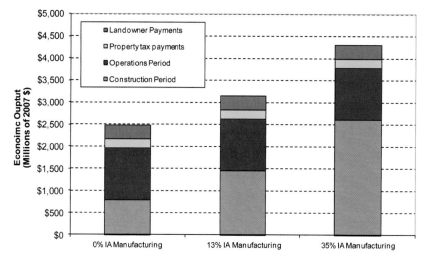

the substantial impact that local manufacturing brings to economic development. In the hypothetical scenario shown, where 2,400 MW of new wind power capacity is installed in Iowa, acquiring 35 percent of the turbines from Iowa-based manufacturers increases the total economic benefit by 55 percent (Lantz and Tegen 2008).

THE ROLE OF PLANNING

Sustainable energy needs, global climate change, and air and water pollution are just a few of the issues challenging today's communities. Planners—tasked with seeing the big picture and thinking about the long term—are integral players in addressing both economic competitiveness and environmental protection.

The most effective energy-policy recommendations facilitate progressive citizen-level actions, as well as decisions, regulations, and land-use plans that direct energy markets toward competitive, healthy, and safe practices. The planning profession already recognizes the urgency of such issues as urban sprawl, the degeneration of inner-ring suburbs, and the disappearance of agricultural and open-space land resources. To create truly sustainable communities, planners must guide stakeholders and communities toward increasing energy conservation and renewable energy production while significantly reducing the use of nonrenewable energy sources. Planners can help stakeholders understand the nexus between, on the one hand, today's energy production and consumption and, on the other, future environmental conditions, economic health, and quality of life.

While planners exert only minimal influence on the selection of energy sources, they can influence energy demand and facilitate new infrastructure development in their communities. Areas of opportunity include the siting of energy generation and transmission facilities, the use of renewable energy, natural resource extraction practices, transportation infrastructure design, resource conservation, industrial development, waste management, and site design. In addition, planners can advance the adoption of resource sustainability principles through comprehensive planning processes, as well as help communities reduce the environmental impacts of electric generation and consumption. In some instances, planners can also influence local energy decisions through the use of subsidies and education.

GREENSBURG, KANSAS

The City of Greensburg, Kansas, has moved to the forefront of sustainable community building in recent years. In 2007, an EF5 tornado leveled 95 percent of Greensburg and displaced more than 500 residents, cutting its already small population of 1,500 to 950. During the rebuilding process, the community decided to reinvent itself as America's greenest community. With the planning help of FEMA's Long-Term Community Recovery (LTCR) division, green-architecture firm BNIM & Associates, and the National Renewable Energy Laboratory (NREL), the community adopted goals related to building LEED-certified civic buildings, developing alternative energy sources, and more. Today, Greensburg has more LEED Platinum–rated buildings per capita than anywhere else in the world, and boasts renewable energy production from wind, solar, and geothermal sources.

The city's sustainability goals were spelled out in the Greensburg and Kiowa County Long-Term Community Recovery Plan, adopted in 2007. As part of its sustainable development component, the plan called for Greensburg to "identify and utilize energy alternatives." The plan further called for the city to "identify city-wide energy generation options" and "create community opportunities for renewable resources." Other planning-related documents, including the 2008 Greensburg Sustainable Comprehensive Plan and Vision Plan, emphasized these policies.

Although discussions of wind farms had taken place before the tornado, nothing had been seriously considered. Under the new plans, development of a wind farm became central to the city's

Source: City of Greensburg

sustainability initiatives. Plans for the farm were discussed in city council meetings, which turned into public hearings. Although Greensburg has siting regulations and ordinances for turbines within the city, the wind farm is located four miles outside of the city, where turbines are allowed by right.

A number of stakeholder groups collaborated on the project. USDA Rural Development provided project funding through a rural infrastructure grant for the city, and the farm was originally developed, owned, and operated by John Deere Renewables. All of the energy produced by the wind farm is purchased by the Kansas Power Pool, which distributes it back to Greensburg and 31 additional cities. In 2010, the farm was purchased in full by the Exelon Corporation.

The Greensburg Wind Farm has an average wind speed of 18 mph with 10 1.25-MW turbines. This is enough to power 4,000 homes, which exceeds Greensburg's needs. About one-third of the energy credits are donated to the city, and the rest are sold by carbon offset–provider NativeEnergy to charter groups including Ben & Jerry's, Clif Bar, Green Mountain Coffee Roasters, Stonyfield Farm, and the Kansas Power Pool.

Adopted in March 2011, the city's new Sustainable Land Development Code permits small wind energy systems by right in all districts, subject to standards; it requires setbacks equal to total system height from public rights-of-way and property lines, but it does not impose tower height limitations. Looking ahead, the comprehensive plan envisions citywide renewable energy generation as a "new value proposition," attracting new businesses and industries to the city, and calls for leveraging a green vision for economic development by encouraging the development of renewable energy–based businesses.

For more information:

- Greensburg Long-Term Community Recovery Plan (2007). Available at www.greensburgks.org/residents/recovery-planning/long-term-community-recovery-plan.

- Greensburg Sustainable Comprehensive Master Plan (2008). Energy; Future Land Use and Policy. Available at www.greensburgks.org/residents/recovery-planning/sustainable-comprehensive-master-plan.

- Greensburg Sustainable Land Development Code (2011). Article 4, Alternative Energy Systems; Section 4.2, Wind Energy Systems. Available at www.greensburgks.org/government/permits-regulations/greensburg-sustainable-land-development-codeview.

KITTITAS COUNTY, WASHINGTON

Kittitas County is a 2,315-square-mile rural county in the center of Washington State. Stretching from the Cascade Mountains to the desert and bounded on its eastern side by the Columbia River, the county is known for its winds. It also hosts transmission infrastructure that carries hydroelectric power from the river, increasing its suitability for large-scale wind energy projects. Four such projects have been built in the county, totaling more than 660 MW and 361 turbines, and a fifth has been approved. Though residents largely support wind power, recent overrides of county siting decisions by the state has caused some controversy.

Wind energy developers first approached the county with a utility-scale development proposal in 2003. Kittitas County's land-use code included a utilities section but nothing that specifically addressed wind energy facilities, so staff developed a Wind Farm Resource Overlay Zone ordinance that created a streamlined permitting process for this use in certain contexts. Dan Valoff, Kittitas County staff planner, explains that the ordinance specifies

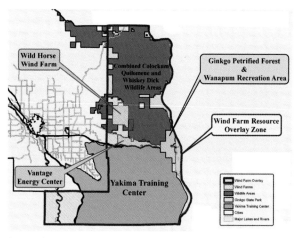

Sources: Washington State Department of Fish and Wildlife; Washington State Department of Natural Resources; Kittitas County Code Title 17.61A; Kittitas County GIS Zoning Data Set; Vantage Wind Power LLC Development Agreement Submittals

the arid, mountainous, sparsely populated areas in the eastern and southern parts of the county as preferred locations for large-scale wind energy. For projects in those locations, the ordinance simply requires developers to go through the county's standard development agreement process, which entails one public hearing in front of the board of commissioners and environmental review. To site a wind farm in other locations within the county, however, developers must obtain a comprehensive plan amendment and rezoning for the parcel before beginning the permitting process. The county has approved two wind farms within the overlay zone and has not approved any projects proposed for lands outside those areas. The state, however, has.

Washington State is one of a handful of states that to some extent preempt local control of wind energy project siting. Developers may choose to bypass local jurisdiction and request project certification from the state's Energy Facility Site

Evaluation Council (EFSEC), which prepares reports on proposed projects and makes recommendations to the governor, who then may approve, reject, or order reconsideration of projects. Though the process takes into consideration local regulations and community input, the results do not always coincide with local wishes. When in 2007 the governor approved a large wind energy project initially denied by the Kittitas County Board of Commissioners, local opponents filed suit, and in 2008 the state supreme court upheld the governor's approval, affirming EFSEC's ability to preempt local authority in this area (Residents Opposed to Kittitas Turbines v. State Energy Facility Site Evaluation Council, 197 P.3d 1153).

Two additional developers whose initial proposals outside the designated overlay zone were denied by the county have since gone straight to EFSEC for approvals rather than negotiate further with the county. Kirk Holmes, director of public works for the county, notes that local governments can provide input to EFSEC—they can petition EFSEC to change turbine locations, for example—and he reports that EFSEC is fairly responsive: "The spirit and intent of the state law is not to skirt local building codes and environmental laws." However, review at the state level may not capture all local concerns and issues.

Once a wind energy project is approved, the developers must meet county development requirements and obtain local permits. Dan Davis, former plans examiner for Kittitas County, emphasizes the importance of the preapplication process, which allows a developer to start meeting as soon as possible with county staff: "This gets all the players at the table—fire, public works, planners, public health—to discuss the permitting issues. The developer then leaves with a list of items needed for building permit approval." Development or staffing agreements are required for the turbines, turbine footings, road building, mechanical equipment buildings, and any impacts on county infrastructure, such as roads. Davis recommends that local governments be proactive in drawing up permit submittal requirements ahead of time. "When developers get the green light, it's typically been a very long approval process and they come in ready to go and breathing down the neck of the building department—but often they won't have all the required documentation, and it can take a significant amount of time to obtain all that information. If the building department has all the requirements spelled out ahead of time and they've communicated that in the preapplication process, this can save a lot of time and money. It helps both sides."

Planners coordinate negotiations between the developer and various county departments: for example, the public works department works with contractors to rate roads before and after turbine construction, with contractors responsible for repairing any damage they cause, and the building department deals with plan review, permitting, and inspections. The public health department enforces water and septic systems requirements for any operations or maintenance buildings constructed. Planners also

(continued on page 7)

(continued from page 6)

oversee environmental review, which includes archaeological and cultural resource surveys; monitoring of wildlife impacts; construction and stormwater discharge BMP requirements; rare plant protections and noxious weed abatement; and fire protection plans. The developer must obtain approvals from state agencies, including the Department of Ecology for stormwater permits and the Department of Fish and Wildlife, which has created bird and bat impact standards for eastern Washington and is involved with continuing data collection. The Yakima Indian Nation is also involved in the archaeological review. The county is not involved in developer negotiations with individual landowners to lease land for projects, however.

Davis notes that Kittitas County has found it more efficient and effective to issue building permits by tax parcels rather than for individual turbines, since utility wind projects can include more than 100 turbines. Where multiple turbines are sited on one parcel, the county gives one building permit for the parcel. This cuts down on paperwork and links the permit to the parcel.

One challenge the county has encountered is working with out-of-state contractors who are unfamiliar with local conditions and regulations. For example, Washington's critical areas legislation provides specific protections to environmentally sensitive features; out-of-state contractors may not be aware of these regulations. Holmes emphasizes that local agencies need to be prepared; he recommends setting up weekly construction meetings and monitoring protocols to keep track of project activity. It is important to be proactive in other areas as well. Kittitas County is especially concerned with reviewing the structural engineering of turbine towers and foundations for project safety, and it includes special inspection requirements up front in its development agreements.

The county also permits small wind energy conversion systems (WECS) for accessory onsite power generation by right in all zoning districts; small WECS are popular with home owners on farmsteads and ranchettes—and, Valoff adds, with university professors. Initially the county code required a building permit with no additional standards provided. With the number of applications growing, the county decided to create specific standards for this use, and Davis developed a small wind ordinance to simplify the permitting process for his department. The resulting ordinance establishes a user-friendly, over-the-counter process for small WECS.

Davis laid out fairly specific application requirements: applicants must submit a site plan, turbine description, and engineering analysis addressing the tower, the tower foundation, and the connection of the tower to the foundation. Davis explains, "This is a new technology, so we need to ensure safety through specific requirements and make sure all projects have appropriate engineering." However, the county accommodates contractors who specialize in small WECS installations; if contractors submit designs engineered for worst-case development scenarios for the whole county, they may then use them as blanket designs for subsequent installations, saving them the additional analyses. Standards limit turbine height to 120 feet and impose a setback requirement of 1.2 times the height of the turbine.

There are currently no community wind facilities in the county; a proposed community wind ordinance that would have allowed landowners to form consortiums to build large turbines was voted down in 2010 due to concerns over visual impacts of the large-scale turbines. In general, however, Kittitas County residents support wind energy. The small turbines are very popular for personal use, and for the most part the community appreciates the importance of the utility-scale projects. However, aesthetics are an issue for some, as the placement of some turbines has marred the county's mountain vistas. Valoff explains, "People like the turbines in the distance in the preferred areas, where they are far enough away to look good. With the state-approved projects, though, some of the large turbines are really in people's faces, and people have concerns." He adds, "It's interesting to work on these projects conceptually, but you really need to see the turbines being built to understand how huge they really are—you need to see it to believe it. It's a very large-scale process and a very industrial use with the maintenance and operations structures and substations needed. It's important to site large turbines in rural areas with little population because of this." Holmes agrees: "Siting these projects is one thing, but construction is another and there can be significant impacts." He encourages local staff that lack wind energy permitting expertise to seek information and advice from other agencies as necessary to ensure the best possible project outcomes.

For more information:

- Kittitas County Zoning Ordinance, Chapter 17.61A, Wind Farm Resource Overlay Zone, and Chapter 17.61B, Small Wind Energy Systems. Available at www.co.kittitas.wa.us/boc/countycode/title17.asp.

- Kittitas County Community Development Services, "Wind Farm Siting Application (for proposing a wind farms in the Wind Farm Resource Overlay zone, as provide[d] for in KCC 17.61A)." Available at www.co.kittitas.wa.us/cds/forms/Wind-Farm-Siting-Application.pdf.

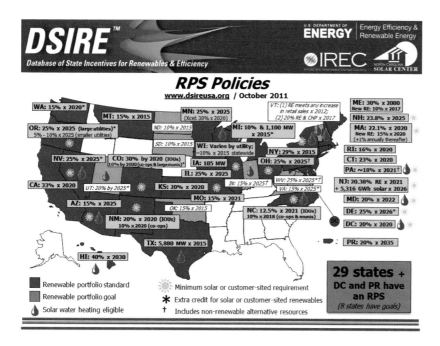

Figure 1.5. *Renewable portfolio standards in U.S. states as of October 2011*

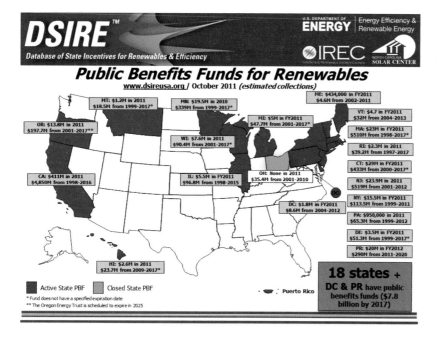

Figure 1.6. *Public benefits funds in U.S. states as of October 2011*

STATE POLICY AND GOALS

Many states have worked to provide an increasingly solid foundation for renewable energy. The renewable portfolio standard (RPS) is the most effective and most popular of state policies for the deployment of renewables. An RPS requires a state's utilities to include a certain percentage of renewable energy in their generation portfolios by a given year (Figure 1.5). These requirements can be instituted by the state legislature (as in Ohio), the public utility regulatory commission (as in New York), or through a ballot initiative (as in Colorado). In addition to a timetable, the requirements usually specify penalties for noncompliance. Depending on the state, public power utilities (i.e., co-ops and municipal utilities) may or may not be required to comply. Because wind energy is usually the renewable generation source with the lowest wholesale cost, it often dominates the RPS portfolio.

Additionally, many states have established funds to promote efficiency and the development of renewable technologies and related projects. Often these funds focus on distributed generation, including wind (Figure 1.6). Some states offer sales tax exemptions or income tax credits, rebates, financing subsidies, and net metering (which allows a power producer to receive credit for electricity generated from an on-site renewable source such as a residential wind turbine). States may also have policies in place (e.g., sales tax exemptions, state production tax credits) to support development of community wind projects. Community wind projects are generally comparable to utility-scale projects except that they include some form of local ownership. They may be multimegawatt or simply one- or two-turbine projects that serve local demand. In some cases, community wind projects may utilize smaller 100-kW turbines or comparable medium-scale machines.

THE FUTURE OF WIND ENERGY AND THE NEED FOR STRONGER LOCAL POLICY

In 2006, President George W. Bush stated that the United States could produce up to 20 percent of its electricity from wind, though at the time it accounted for less than 1 percent. A collaboration among the U.S. Department of Energy, several national labs (led by the National Renewable Energy Laboratory), and the wind industry (including the American Wind Energy Association) was subsequently tasked with identifying more specifically what a "20 percent wind future" might entail, including quantifying the benefits and identifying the challenges, and how it might be achieved (DOE 2008). The team explored seven key areas: wind system technology, manufacturing and resources, transmission and grid integration, siting and environmental effects, markets, policy, and benefits. It concluded that the 20 percent goal was attainable by 2030 without any dramatic technology breakthroughs and would provide many local and national benefits and efficiencies. This would, however, require significant deviation from business as usual.

The study scenario, along with the Energy Information Administration's estimate of electricity demand in 2030, indicated that a 20 percent wind future would require 305,000 MW of wind energy, including 54,000 MW from offshore wind-power plants. The analysis indicated that there was as much as 600,000 MW of developable wind resources at a cost of $40 to $60 per megawatt-hour. Forty-six states would have substantial wind development by 2030, including 35 that would have more than 1,000 MW installed (Figure 1.7). The total footprint of land required for wind energy projects was estimated at about 50,000 square kilometers, or about 80 percent of the size of West Virginia. Turbines, service roads, and related equipment would require between 1,000 and 2,500 square kilometers, or less than the area of Rhode Island. This scenario would also require building 12,000 to 15,000 miles of new high-voltage transmission lines (DOE 2008).

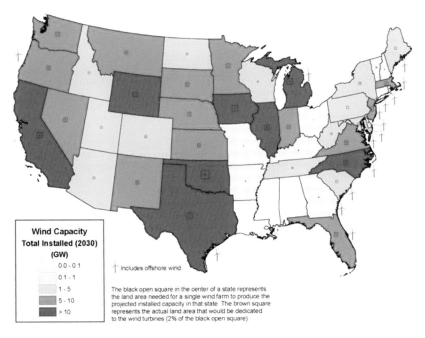

Wind Capacity Total Installed (2030) (GW)
0.0 - 0.1
0.1 - 1
1 - 5
5 - 10
> 10

↑ Includes offshore wind

The black open square in the center of a state represents the land area needed for a single wind farm to produce the projected installed capacity in that state. The brown square represents the actual land area that would be dedicated to the wind turbines (2% of the black open square).

Figure 1.7. Distribution of installed wind power capacity (GW) among states under the 20% Wind by 2030 scenario
Source: DOE 2008

In addition to addressing transmission infrastructure needs, other challenges include developing and enabling federal and state policies, further developing U.S. manufacturing and human resources, streamlining siting and permitting processes, implementing modest technology improvements, improving utility coordination and systemwide operational practices, and increasing social acceptance of wind facilities (DOE 2008).

This PAS Report supports the aim of "20% Wind by 2030" by providing planners with the tools and strategies they need to help plan for, open, and responsibly develop wind energy markets. Planners are uniquely positioned to strengthen local wind energy policies by addressing market barriers at strategic points of intervention while also ensuring that siting and development standards for wind energy facilities and equipment are consistent with local community expectations. This report is intended to help community planners effect policy and regulatory change, build stakeholder support, and provide key technical information to public officials.

CHAPTER 2

Industry Overview

Ruth Baranowski with Eric Lantz

 In 2010, the U.S. wind energy industry installed just over 5,100 MW. Although comparable to installations in 2007 and well above installations prior to 2007, installations in 2010 were down nearly 50 percent from those in 2009 and roughly 40 percent from 2008 (Wiser and Bolinger 2011). This was the first year-to-year reduction in installations since 2003–2004. Nevertheless, installations in 2010 did increase the country's cumulative installed capacity by 15 percent. At the end of the year, total U.S. wind power capacity exceeded 40,100 MW (Wiser and Bolinger 2011).

The U.S. wind resource covers much of the country (Figure 2.1). The wind resource potential is greatest over much of the Great Plains, but wind development is occurring throughout much of the continental United States. (See Figure 1.3.) Development is affected by state policy, regional markets (i.e., existing wholesale generation assets, demand for additional power capacity, and the mix of current regional power capacity), the wind resource, transmission access, and other factors.

***Figure 2.1.** The contiguous U.S. wind resource 80 meters above ground level*

Source: NREL and AWS Truepower

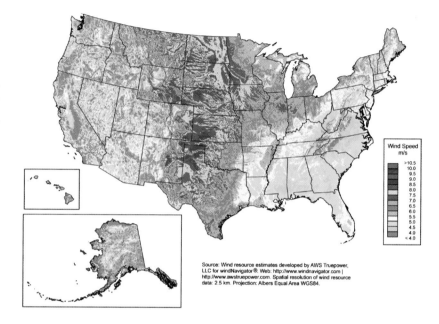

Source: Wind resource estimates developed by AWS Truepower, LLC for windNavigator®. Web: http://www.windnavigator.com | http://www.awstruepower.com. Spatial resolution of wind resource data: 2.5 km. Projection: Albers Equal Area WGS84.

The leading states in terms of installed capacity include Texas, Iowa, California, Minnesota, Washington, and Oregon (Wiser and Bolinger 2011). In 2010, Texas became the first state to surpass 10,000 MW, with a total of 10,085 MW of wind capacity installed at year-end. Also in 2010, the number of states with utility-size wind turbine installations increased to 38 following the addition of Delaware and Maryland. (See Figure 1.3, page 3.) Through 2010, 14 states had more than 1,000 MW installed.

The wind industry consists of a diverse and multifaceted array of applications and business models.

DISTRIBUTED (SMALL) WIND

Distributed wind energy systems are sometimes referred to as residential or small wind systems and typically consist of turbines with capacities up to 100 kW. Distributed wind turbines are often installed "behind the meter"— meaning they serve on-site consumption and are not designed to sell power to a utility or wholesale power purchaser. Normally mounted on towers or poles, small turbines can be used to charge batteries and to power remote telecommunications sites and villages. They also power homes, businesses, and isolated water pumps that feed irrigation or livestock watering tanks. Small wind turbines may be installed alone as a primary source of power or as part of a hybrid system. They are often net-metered. (See Chapter 1.) Historically, specialized distributors and installers linked small wind–turbine manufacturers to consumers. However, today manufacturers may market directly to consumers, and some turbines are now available at large-scale retailers such as Lowe's and The Home Depot.

Unlike the broader industry, the small wind industry experienced continued growth in 2010, up 26 percent. Table 2.1 (page 14) shows that nearly

CITY OF RENO AND WASHOE COUNTY, NEVADA

Reno, Nevada, population 219,636, is located in Washoe County in the high desert at the foot of the Sierra Nevadas. Nevada has abundant solar and wind resources, and the western part of the state experiences the "Washoe Zephyr," an afternoon wind, most summer and midwinter days.

Both the city and the county became interested in developing wind energy regulations in the early 2000s. With several active installers in the community, Reno had been processing permits for small wind energy conversion systems (WECS) since the late 1990s. After many years of requiring special use permits for small WECS, in 2008 the City adopted an ordinance making small WECS an accessory use permitted by right in all districts subject to standards. A building permit is required; applications for wind turbines must include standard structural drawings, engineering analyses, and line drawings of electrical components. The ordinance imposes setbacks of 30 feet from front property lines and 10 feet from side and rear property lines, sets noise limits, requires nonreflective, nonobtrusive colors, and allows for combined uses with structures such as communication structures or flagpoles. There are no height limits, and turbines are exempted from utility screening requirements. As required by state law, home owners association (HOA) approval is required where applicable, though there are few HOAs in Reno. Large wind energy systems are not addressed in Reno's code.

In contrast, Washoe County received fewer permit applications than Reno until recent years and had no WECS standards in place when in 2008 it received a proposal for a 44 MW wind farm on private land. The proposal was approved in 2009 over significant public opposition; the Virginia Peak wind farm will be the first in the county when it is built.

In 2010, the county amended its zoning ordinance to add standards for both large and small WECS. The ordinance defines "private wind machines" as those with rated capacities of 100 kW or less, and only one may be installed on parcels smaller than one acre. Any WECS with a rated capacity over 25 kW or greater than 75 feet tall on a parcel under five acres (100 feet in height if the parcel is greater than five acres) requires a special use permit. The ordinance sets a minimum setback equal to the overall turbine height, but this may be reduced with written consent of abutting property owners. It also sets noise, aesthetic, and safety requirements for small turbines. Commercial wind machines are considered principal uses and are permitted as special uses in certain rural and industrial districts. Permit application requirements are comprehensive and include site plans, regrading and revegetation plans for temporary construction roads, drainage and erosion control plans, and FAA approvals. The ordinance also establishes setbacks, noise and height limits, and aesthetic standards for commercial wind machines; additional requirements include postconstruction noise compliance studies, postconstruction wildlife injury and mortality monitoring plans, and decommissioning performance securities and plans.

Nevada state law requires that zoning regulations be designed to promote the use of solar and wind energy, and it bars local governments from adopting any regulations that prohibit or unreasonably restrict property owners from using WECS. Because this limits the ability of local governments to regulate many aspects of wind turbines, the regulations adopted by Reno and Washoe are more favorable for WECS than those in some communities in other states. Staff in both communities reported very little controversy around the adoption of the regulations and suggested a likely reason was that concerned parties quickly realized the limits of local control.

In Reno, wind energy enjoys solid public support, in part due to the wind resources in the area and in part because of the community's independent streak. In fact, staff expected more opposition to the ordinance than they saw; the idea of wind turbines has been very well received. The City also sent its permitting and inspection staff to a class to train them in evaluating turbine applications and built projects.

Reno city staff estimate that there are between 50 and 100 small wind turbines operating within the city limits, with few known problems or complaints. The City has developed extensive outreach programs for alternative energy and has installed nine WECS, including two vertical turbines, at four city facilities as energy-saving and demonstration projects. In early 2011, the city launched a "green energy dashboard" (http://greenenergy .reno.gov/energy) that shows the energy generated from demonstration wind and solar projects in real time, as well as a wind resource map showing real-time and historical wind measurements from 30 points in and around Reno.

In contrast, according to Washoe County planning staff, the issue of height and resulting visual impact has been a challenge in adopting WECS regulations there, and wind energy continues to be controversial in parts of the county. Recently, several neighbors brought a lawsuit in the district court against the owner of a 75-foot WECS, approved and installed under county regulations (*Forest Hill Subdivision vs. Sowers*, case CV11-00080). The neighbors alleged that the WECS was a nuisance based on aesthetic impact alone. A county staffer was called in as an expert witness. The judge ruled in favor of the neighbors, and it remains to be seen if the WECS owner will appeal.

Compared to Washoe County, Reno has had more and taller WECS for a longer period of time, and yet has less controversy over their impacts. There are many possible explanations for this. Perhaps different educational efforts or community values about the importance of alternative energy generation explain the difference. Perhaps in the relatively rural county, WECS are experienced as more visually disturbing than they are in the more visually cluttered city. Another possibility is that the taller WECS allowed in Reno actually create less visual impact to observers on the ground, as they are farther from view. Finally, Reno residents may have grown accustomed to seeing WECS and thus

(continued on page 14)

(continued from page 13)

are not as bothered by them as are county residents, where WECS are a newer part of the landscape. Any combination of these factors—or some other one—may explain the different attitudes toward WECS in these two coterminous communities.

For more information:

• City of Reno Land Development Code (2011). Chapter 18.08, Article II, Section 18.08.203(e)(6)(d), Standards for Specific Accessory Uses – Utilities, Alternative Systems. Available at http://library.municode.com/index.aspx?clientID=14345&stateID=28&statename=Nevada.

• Washoe County Development Code (2011). Division 3, Article 326, Wind Machines. Available at www.co.washoe.nv.us/comdev/publications_maps_products/comdevcode/comdevcode_index.htm. ◀

TABLE 2.1. ANNUAL SALES OF SMALL WIND TURBINES IN THE UNITED STATES

Year	Number of Turbines	Capacity Additions (MW)	Sales Revenue (millions)
2005	4,324	3.3	10
2006	8,329	8.6	33
2007	9,092	9.7	42
2008	10,386	17.4	73
2009	9,800	20.3	82
2010	8,000	25.6	139

Source: AWEA

8,000 units were sold, constituting more than 25 MW of capacity with a value in excess of $139 million. Cumulative small wind installations now exceed 175 MW. Though total capacity continues to grow, the number of turbines sold annually has decreased since 2008, reflecting a shift toward larger turbines. The United States has a robust small wind manufacturing industry, with 2010 sales capturing more than 80 percent of the market and seven manufacturers reporting sales in excess of 1 MW (AWEA 2011b).

The American Wind Energy Association (AWEA) estimates that costs for small wind turbines average roughly $5,500 per kW. At these costs, a typical 5-kW turbine may cost between $25,000 and $30,000 (AWEA 2011b). However, individual costs vary widely due to site-specific factors such as zoning and permitting costs and interconnection fees. The payback period (the time during which the savings resulting from a system equals the cost of installing it) can be as much as 30 years, depending on available incentives and the quality of the wind resource; however, well-sited small wind turbines with incentives can pay for themselves within 15 years (Forsythe et al. 2000). Because of these long payback periods, policy support at the state and federal levels remains critical to the viability of the small wind market segment.

Small wind turbines for the urban environment, sometimes known as built-environment wind turbines, have recently garnered some attention. Often these types of machines are building-integrated, building-mounted, or ground-mounted systems used to offset energy costs or to display environmental commitment. Special engineering reviews are required for building integration in order to minimize structural impacts and maximize turbine performance. Built-environment wind turbines tend to experience higher turbulence than nonurban systems, which can significantly affect turbine production and long-term reliability and should not be dismissed lightly. Built-environment turbines are largely expected to be a niche within the small wind sector as suitable locations with a valued wind resource and viable installation sites are extremely difficult to find. Few jurisdictions have zoning ordinances that cover this type of application, but planners should be aware that proposals for built-environment wind turbines may arise.

In an attempt to ensure that the small wind industry will continue to develop sustainably, the industry has initiated several activities. In 2006, a number of manufacturers and other interested parties formed the Small Wind Certification Council (SWCC), which certifies small wind turbines, as a way to build consumer confidence. In 2010, the North American Board of Certified Energy Practitioners (NABCEP) established a small wind–installer certification exam to ensure that installers can successfully and safely mount any small wind electric system. The goal of this undertaking is to further ensure safety, quality, and consumer acceptance of small wind installations throughout the United States (AWEA 2011b). The Distributed Wind Energy Association (DWEA) was also formed. In further support of consumers, articles for the National Electric Code 2011 were created; some have revised language specifically for small wind turbine installations. Parts of the new articles were modeled after existing language pertaining to solar photovoltaic systems.

Although distributed wind has grown substantially over the five years (Table 2.1), this market segment faces a number of challenges. Along with cost, other concerns also impacting the market include:

- Zoning restrictions (especially for structure heights)

- Aesthetic concerns from neighbors

- Noise concerns

- Proper assessment of wind resources to allow turbines to achieve power production potential

- Warranties (small companies may not provide warranties, and lenders may require them)

- Availability of technicians

- Availability of spare parts

In addition, the broader economic slowdown coupled with fragile state and local policy incentives for small wind has reduced demand and introduced uncertainty into the distributed wind market.

MIDSIZED WIND TURBINES

Midsized turbines are used at schools, farms, factories, private and public facilities, remote locations, and on tribal lands to generate electricity. The size of these turbines (100 kW to 1 MW) often allows them to be installed where the electricity is to be used, thus minimizing the need for new electric transmission lines. Like distributed (small)

FEDERAL TAX POLICY FOR WIND ENERGY DEVELOPMENT

Larry Flowers and Eric Lantz

Wind energy's benefits drive both federal and state policies supporting wind energy development. In 1992, Congress attempted to level the playing field between conventional generation technologies and renewables by creating a production tax credit (PTC) for commercial-scale renewables projects, including wind. The PTC provided wind installations with a 10-year credit of $0.015 per kilowatt-hour (adjusted for inflation) but had restrictive conditions on the qualifying income against which the credit could be applied. In general, large, for-profit corporations are in the best position to monetize the value of the tax credits; individual investors are typically poorly positioned and as a result, have been largely unable to participate in financing for wind projects. In addition, this incentive was not made a permanent part of the tax code. The PTC requires periodic reauthorization by an act of Congress and has expired on three occasions. A series of lapses and 12- to 24-month extensions in the early to mid-2000s generated significant uncertainty in the wind industry and resulted in multiple boom-and-bust cycles, with the effect of discouraging long-term investment (Wiser et al. 2007). More recent two- and three-year extensions beginning in 2005 have supported significant new investment in wind component manufacturing in the United States. By the end of 2010, the United States was estimated to have 400 wind-material manufacturing facilities in 42 states, supporting 20,000 related jobs (AWEA 2011b).

With the financial crash in 2008, the PTC lost much of its value as a useful incentive as profits evaporated and liabilities against which the credit could be used disappeared or were greatly diminished. In order to maintain industry momentum, policy makers provided investors with the option of accessing a direct one-time grant payment from the U.S. Treasury via the 30 percent investment tax credit (ITC), an incentive roughly comparable in value to the PTC but based on initial capital expenditures rather than plant production. To receive the grant, wind projects must first elect to take a 30 percent investment tax credit (ITC) and then convert it to a direct payment from the Treasury. Wind projects can continue to elect the 30 percent ITC through 2012, but the direct grant program expires at the end of 2011. Historically, wind projects have not had access to the 30 percent ITC, which allows the project owner a tax credit roughly equivalent to 30 percent of the capital expenditures associated with the project. The ITC and its grant counterpart are generally more advantageous for low wind-speed sites, where power production per dollar invested is lower. In contrast, the PTC, which provides a tax credit based on actual production for a period of 10 years, requires a much greater period of time to realize the value of the tax credits and entails production risk (i.e., if the facility doesn't produce, no tax credits are generated).

An additional federal policy incentive for which wind projects are eligible is the Modified Accelerated Cost Reduction System (MACRS), which allows capital-intensive wind projects to fully depreciate their value in five years.

Distributed wind projects (up to 100 kW) are typically not eligible for the PTC, which requires sale of the electricity to an unrelated third party. However, they are eligible for the 30 percent ITC. There are, in actuality, two different ITCs which have slightly different rules for distributed wind. The business energy ITC has similar rules around qualifying income as the PTC. The residential renewable energy tax credit does not have the same restrictive rules limiting the types of income that can qualify for the tax credit.

Distributed wind projects are owned primarily by small businesses, nonprofit institutions (e.g., schools), and individual consumers. The ITC for distributed wind extends to 2016. Another incentive for distributed wind is the Department of Agriculture's Rural Energy for America Program (REAP), which started in 2002. The purpose of REAP is to encourage agricultural producers and rural businesses to invest in renewable energy, both to save money and to sell excess production to a local utility. REAP provides both grants and guaranteed loans.

wind, they may be a part of a stand-alone power system or combined with other on-site generation (e.g., diesel) in a hybrid system. Midsized turbines are able to net meter in some cases; however, they often exceed net-metering size restrictions.

The midsized turbine market has struggled due to limited turbine availability and unfavorable project economics. Midsized projects represent significant investments, potentially as much as $1–$2 million. They may have difficulty capturing the economies of scale associated with larger utility-scale projects, and they generally cannot completely offset higher-cost retail electricity consumption, as is often the case with distributed wind. In response to these barriers, in 2010 the U.S. Department of Energy (DOE) designated up to $6 million to advance midsized wind turbine technology, with the intent of boosting the speed and scale of deployment in this market segment. Along with the primary barriers noted above, a host of additional challenges also affect the midsized market sector, including:

- Significant equipment lead times

- Limited proof of technical viability and standard warranties, which are often required to finance projects

- Availability of technicians

- Availability of spare parts

- Lack of regulatory support/consideration

- Lack of standardized interconnection policies and procedures

- Permitting/siting challenges

UTILITY-SCALE TURBINES

The U.S. wind industry is dominated by the utility-scale market segment. This segment produces power for sale into wholesale power markets around the country. Utility-scale turbines are large. Today, typical hub heights are 80 meters (about 262 feet), with rotor diameters ranging from 80 meters to more than 100 meters. Typical capacities range from 1.5 to 3 MW. Utility-scale turbines have increased in size substantially since 2000, when hub heights and rotor diameters were between 50 and 60 meters and average power was less than 1 MW (Wiser and Bolinger 2011).

Since 2006, the utility-scale sector in the United States has grown at a compound average rate of more than 28 percent. Moreover, with the exception of 2010, wind power has been the second-largest resource (by capacity) added to the U.S. electricity grid since 2005. This period has also seen the development of a robust domestic manufacturing sector. In 2006, domestic content in the U.S. wind industry was estimated at 35 percent; today it is on the order of 60 percent (Wiser and Bolinger 2011).

The outlook for the utility-scale sector is somewhat mixed. Looming policy expirations for the PTC, ITC, and Treasury grant programs are expected to drive significant installations in 2011 and 2012, but installations could fall again in 2013 (see sidebar, page 15; Wiser and Bolinger 2011). The industry also faces competitive pressures from other power generation resources—namely, natural gas—and has been affected by the overall downturn in the U.S. economy, which has reduced demand for power throughout much of the country. At the same time, however, utility-scale wind costs are falling. While capital costs increased steadily from 2004 to 2010, turbine prices have come down 20 to 30 percent from their peak in 2008, and installed project costs are also beginning to retreat from their 2010 highs (Wiser and Bolinger 2011). (There is a lag time between changes in turbine prices and changes in project capital costs, because of the

▶ **GRATIOT COUNTY, MICHIGAN**

Michigan's Gratiot County, population 42,476, has become a trailblazer in regional planning in one of the nation's strongest home-rule states. As a result, large-scale wind energy developers have been beating a steady path to its door, and the revenues promised by future wind farm development may help lift the county out of one of the worst economic recessions to hit Michigan since the Great Depression. Led by local economic development organization Greater Gratiot Development and the management and elected leadership of the County and the municipalities within its boundaries, the Gratiot community has been a vanguard in preparing itself for large wind development through planning.

Gratiot County is located in the center of the state, with more than three-quarters of its lands in cropland, pasture, or forest. The county was once a hub for heavy industry, but in the 1970s the loss of several major employers and changes to the automobile industry and its supplier chain sent unemployment skyrocketing to 21 percent. In response, Gratiot County and its constituent municipalities worked with the private sector to incorporate Greater Gratiot Development, a nonprofit organization to coordinate economic development and related services throughout the county. Thus began a tradition of regional collaboration and planning that would serve the county well.

In 2008, motivated by the worsening economic recession, Greater Gratiot Development, under the direction of its president, Don Schurr, obtained funding to involve the County and 21 municipalities in a countywide master planning effort. This would eventually result in 2010's Gratiot Regional Excellence and Transformation (GREAT) Plan, the first countywide, locally developed, shared, adopted, and implemented plan in Michigan. Around the same time, the wind industry had begun showing interest in the county as a potential site for development. The developer Wind Resources was attracted by the existing grid network of transmission lines in place from the county's legacy of heavy industry and oil refining, and its tower tests in 2008 demonstrated sufficient wind capacity for a feasible project. The timing was perfect to integrate the development of wind energy regulations with the communitywide master-planning process.

As economic committee chair of the master plan project, Schurr introduced wind energy as a potential economic development strategy in the beginning of the planning process and suggested the group simultaneously develop a wind energy ordinance to accommodate future wind energy development. Schurr further advocated for full and transparent participation in the development of both: anybody who wanted to participate could do so. Dan Rossman, the local Michigan State University extension director, organized a countywide wind energy education effort with the Gratiot County Farm Bureau for landowning farmers, which culminated in a large meeting with expert presentations on wind energy development. The County retained Spicer Engineering of neighboring Saginaw County to provide technical guidance. In May 2009, the county planning commission unanimously approved a wind energy ordinance applying to six county-zoned townships and serving as a model for other municipalities.

The ordinance focuses solely on large-scale wind energy conversion systems designed to supply energy to off-site customers, and aims to protect landowners and the community while allowing wind farm projects to be developed. It provides for the creation of Wind Energy Facility Overlay Districts (see map) through zoning map amendments, in which wind farms are permitted uses with the granting of a Wind Energy Facilities Permit through the special use permitting process. The ordinance further notes that Wind Energy Facility Overlay Districts are intended as agricultural preservation measures.

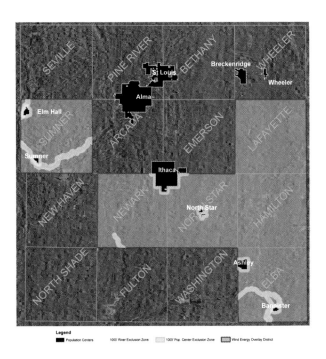

Gratiot County wind energy overlay district map
Source: Gratiot County

Required wind farm permit application materials include a narrative of the proposed wind farm, a site plan, and a decommissioning performance bond of at least $1 million. The proposal must comply with applicable water resource protection, erosion, and wetlands ordinances, and it must mitigate the project's visual appearance through minimal lighting and undergrounding of electrical lines where practicable. Turbines must be set back by the greater of 1,000 feet or two times hub height from residences and occupied buildings and by the greater of 400 feet or 1.5 times hub height from public roads, railroad lines, and "rails to trails" facilities. To minimize disruption to agricultural activity, turbine and access road siting is encouraged along internal property lines, but a setback of 1.5 times hub height applies to property lines of nonparticipating parcels. The standards also restrict shadow flicker to 30 hours per year, restrict noise to 55 dbA at nearest habitable structures, and require safety measures (warning signs, no public or climbing access to turbines). (See Chapter 7.)

(continued on page 18)

(continued from page 17)

Adoption of the wind energy ordinance created a consistent framework for wind energy development, increasing the attractiveness of Gratiot County to wind energy developers who dread grappling with patchworks of differing local regulations. Groundwork for the first wind energy facility permit, sought by Wind Resources in partnership with Invenergy, the country's largest independent wind-energy generation company, was laid by a series of public meetings, including "coffee and cookie" gatherings led by Wind Resources and Invenergy to allow the public and company representatives to discuss wind turbine issues. These meetings culminated in a large public hearing at Breckenridge High School in March 2010 hosted by the four townships in which Invenergy intended to site the proposed wind farm. More than 400 people turned out to discuss passage of the countywide wind energy ordinance and potential impacts such as shadow flicker, future development, property values, noise, wildlife considerations, and impingement on geese flyways. For every voiced concern, others who had lived near wind turbines gave testimony that perceived negative impacts were unjustified. Invenergy staff showed diagrams, models, data, and photographs of the wind farm construction process and wind farms currently in operation. The final result was a unanimous vote by the four townships to approve the special use permit for the project.

Construction for the Invenergy project is currently under way for 133 turbines; the company had already secured a power purchase agreement with DTE Energy, the largest energy provider in Michigan. More than 200 families are part of the leasing pool of Invenergy's project, which includes project neighbors without turbines on their parcels. Individual lease owners will receive proportional percentages of the gross proceeds. The first stage of the project will provide 150 skilled construction jobs, 15 full-time technician jobs, and $1.2 million in annual revenue for the county and municipalities. Landowners will get $80 per acre for leased space and a percentage of gross royalties. The project is expected to generate enough electricity to power 54,000 homes annually. One local official estimates that property taxes generated by the project and royalty payments to lease owners could amount to $100 million over the next 20 years. Breckenridge city manager Jeff Ostrander told the press that his community's schools could capture up to $800,000 in the first year.

The Invenergy project is just the beginning, according to Gratiot County officials. A special permit application for a second wind farm from locally initiated Beebe Wind, in partnership with Nordex of Germany, was approved unanimously in February 2011; the proposed project could encompass up to 100 turbines and produce 300 MW of energy. A third special permit for TradeWind Energy, a Kansas-based firm, has been approved for a project of 150 MW. In addition, Invenergy is now looking at a second project in Gratiot County of approximately 200 to 300 MW.

The primary reasons for Gratiot County's wind energy windfall continue to be a willing community, a viable wind resource, and good access to transmission. The faltering economy in Gratiot County made the public more receptive to countywide planning and wind energy than they might otherwise have been; the vast majority of citizens at public hearings are in favor of bringing turbines to their rural communities. Wind energy is an exciting alternative to the struggling automobile industry and dependence on foreign oil, and one that is highly compatible with the county's agricultural base; farmers appreciate the farmland preservation aspects of the wind district overlay and the chance to diversify their income streams by hosting wind turbines on their properties. Don Schurr makes the point that planning has made the difference: though Gratiot County is not the best place in Michigan for wind, the fact that county communities are all on the same page when it comes to regulating wind energy has helped land them the state's largest wind farm, with more turbines (and more revenue) yet to come.

For more information:

- Gratiot County Adopted Wind Ordinance. Available at www.co.gratiot.mi.us/LinkClick .aspx?fileticket=HV9KF5k0PL4%3d&tabid=176.

- Greater Gratiot Development. "About Us." Available at www.gratiot.org/1/292/index .asp.

- TradeWind Energy. "Gratiot Farms Wind Project." Available at www.tradewindenergy .com/Project.aspx?id=1816. ◄

delay between the negotiation of turbine contracts and power purchase agreements and completion of actual projects.) Technology performance is also continuing to improve; the increases in hub heights and rotor diameters over the past decade are driving up project production (per unit of installed capacity) and reducing the cost of electricity produced by wind turbines. Falling capital costs and improved performance suggest that wind could be an increasingly competitive power generation resource.

In addition to the high-level market drivers of demand for electricity and competition from alternative generating resources, a number of additional barriers to utility-scale wind persist, including:

- Inconsistent federal policy

- Access to transmission

- Concern over impacts to wildlife (including protected and unprotected species)

- Aesthetic and nuisance concerns from potential host communities

Offshore Wind

A subset of the utility-scale wind industry is offshore wind energy development. Although no offshore wind power plants exist in the United States today, there is significant interest in developing such projects. Through 2010, there were more than 2,000 MW in relatively late stage development (Wiser and Bolinger 2011). The offshore wind resource is generally preferable due to higher wind speeds and lower turbulence. With significant power demand located in major metropolitan areas on both coasts (e.g., New York City; Boston; Washington, D.C.; Los Angeles; San Francisco) and a robust coastal wind resource (Figure 2.2), offshore projects also can be placed closer to large demand centers, reducing the

OFFSHORE WIND ENERGY

Kitty Fahey

In the coming months and years, local officials and planners along the coasts and Great Lakes will be increasingly involved in preparing their communities for the economic benefits—and infrastructure complexities—that offshore wind energy facilities can bring.

The National Offshore Wind Strategy, developed by the U.S. Department of Energy (DOE) and Department of the Interior (DOI), has set ambitious goals for this technology—10 gigawatts of offshore wind-generating capacity by 2020, at an energy cost of 10 cents per kilowatt hour. (One gigawatt equals one billion watts.) Officials aim for 54 gigawatts of offshore wind-generating capacity by 2030, at an energy cost of 7 cents per kilowatt-hour.

The offshore areas considered for wind energy development lie on the U.S. outer continental shelf and are held in the public trust—therefore, these areas are not technically within the planning jurisdictions of local coastal communities. However, as the offshore wind energy industry progresses, it will bring great economic opportunities—and also great planning and infrastructure challenges—to onshore communities nearby. For this reason, local officials and planners will want to take part in conversations about

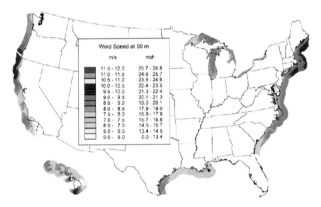

Figure 2.2. *Map of the offshore U.S. wind resource, 90 meters above sea level*

Source: NREL / Schwartz et al. 2010

where, when, and how offshore wind development proceeds.

National interest and activity in the development of offshore wind energy has never been greater. "In the past several years, the federal government has set both a planning framework for offshore areas and a regulatory process for developing renewable energy in federal waters," says Adam Bode, a spatial analyst with The Baldwin Group at the National Oceanic and Atmospheric Administration (NOAA) Coastal Services Center. The center develops tools and guides to assist coastal planners and others with issues related to planning of offshore areas.

Bode adds that 29 states and the District of Columbia have established goals or laws requiring a certain percentage of electricity to be supplied by renewable energy. "Many of these states will find their goals hard to reach without adding renewable offshore wind energy—and, to a lesser degree, wave and current energy," he says.

Local officials and planners are invited to become involved in state-level discussions on offshore wind energy development. The U.S. Department of the Interior's Bureau of Ocean Energy Management (BOEM), which has jurisdiction over offshore wind energy development on the outer continental shelf, is helping to form federal-state task forces that will establish offshore wind energy areas (WEAs).

"When an offshore area is designated a WEA, it means this area has a decreased likelihood of causing conflicts among stakeholders or having 'fatal flaws,'" says Bode. "But designating a WEA is just a starting point—much more feedback and investigation will be needed to determine the very best areas," he says.

"The task forces will aid communication between BOEM and state, local, tribal, and federal stakeholders regarding leasing and development issues on the outer continental shelf. The task force meetings welcome anyone and provide an opening for planners or any other sectors and stakeholders to keep updated on developments, provide information and feedback, and lodge potential objections during the decision-making process." (To learn more about the progress of WEA task forces in various states, visit www.boem.gov/Renewable-Energy-Program/State-Activities/Index.aspx.)

Offshore Wind Requires Planning

Until recently, many coastal and Great Lakes communities were reluctant to discuss offshore wind energy development because it was perceived to be prohibitively expensive, compared with land-based wind energy farms.

One government initiative is working to lessen the obstacles that have stood in the way of offshore wind energy development. In early 2011, the DOE and DOI unveiled a plan to streamline the development of offshore wind energy as part of DOI's "Smart from the Start" initiative (www1.eere.energy.gov/femp/news/news_detail.html?news_id=16709). The initiative makes available more than $50 million in funding opportunities intended to remove market barriers and speed development of next-generation technologies. As part of the initiative, four Mid-Atlantic WEAs off the coasts of Delaware, Maryland, New Jersey, and Virginia will undergo accelerated environmental reviews, with areas off the North Atlantic and South Atlantic coasts soon to follow.

"Offshore wind energy development is more expensive up front, but it's still attractive in some areas because it's close to the power demand centers that are dense in population—New Jersey is one example," says Bode. "There's a great potential for the creation of skilled local jobs in construction, operation, and maintenance of offshore wind facilities. For planners, those jobs will also have also big impacts on local infrastructure planning," he says.

One infrastructure consideration concerns offshore wind towers, which can be 400 feet tall. Deep-draft ports will be needed to support the traffic of large ships that install and service the towers. Another complexity involves the connection to the power grid. "A cable will have to run from the facilities back to shore, where it plugs into the grid. That cable is going to cross federal and state waters and the local jurisdiction. In addition, the community needs to make sure that the grid connection point has the capacity to handle the wattage coming in from the wind farm," adds Bode.

(continued on page 20)

(continued from page 19)

The offshore-wind supply chain industries can be an economic boon to communities, but they will also involve intense infrastructure planning. "For the best logistics and efficiency, the manufacturers of large components would be integrated alongside the ports of deployment, as we see in the European ports today," says Patrick Fullenkamp, the director of technical services for the Great Lakes Wind Network, an international supply-chain advisory group and manufacturer network. "With the larger parts and high cost and limitations to transport [of offshore wind facilities], the need for coastal manufacturing sites for heavy fabrications, casting, forgings, composite blades, [and] nacelle assembly will be required."

Paul Wolff, a councilmember for the City of Tybee Island, Georgia, is one of many local advocates who are involved in making offshore wind energy production a reality. A parent company of Georgia Power is in the process of applying for two offshore leases to test wind resources for a wind farm 10 miles off the Georgia coast.

"With all the developments in offshore wind, coastal communities will need to start considering the offshore environment in their local planning decisions," says Wolff, adding that there are advantages to doing so. "Aside from all the environmental benefits, the offshore wind industry has the potential to be a huge economic driver for an entire region." Wolff cites the United Kingdom, which is committed to expanding its offshore wind resources but lacks a strong supply chain for all the component parts that must be manufactured in the coming decades. "There are many opportunities for U.S. companies and manufacturers, if we pursue them," he says.

But before an offshore wind farm can be approved, a wide variety of community sectors need to take part in the planning and development process—the fishing industry, national security interests, officials leasing submerged lands, coastal community developers, and agencies protecting critical habitat, to mention just a few.

"Here, our meetings involve local residents, organizations, city and county staff, and about a dozen agencies connected with ports, fisheries, natural resources, the U.S. Army Corps of Engineers, and others," says Wolff.

Planning Tools for Offshore Areas

The NOAA Coastal Services Center provides guidance, tools, and data that can ease the learning curve for coastal officials and planners who need to start planning for offshore area uses. Some resources planners can use to help plan for offshore wind energy development include the following:

Multipurpose Marine Cadastre (www.marinecadastre.gov): This screening tool enables users to find authoritative and relevant data and information to refine wind-energy site searches, create and customize detailed maps, and address project ideas with collaborators, regulators, and stakeholders. To see how the North Carolina Wind Energy Task Force is using the cadastre to help identify outer continental shelf lease blocks suitable for wind energy siting, visit http://explorer.arcgis.com/?open=450861e4d71448639fec7055213d7 c03. The cadastre effort is led by the center and the BOEM.

Offshore Renewable Energy Planning Site (www.csc.noaa.gov/digitalcoast/energy): This website features information on many tools, guides, data sets, and trainings that can assist anyone involved in finding the best location for offshore renewable energy projects. A sampling includes CanVis, a visual simulation tool that can be used to illustrate the visual impact of offshore wind turbines; the Benthic Terrain Modeler, which helps users examine the deepwater environment; and a guidebook, *Marine Managed Areas: Best Practices for Boundary Making*.

Legislative Atlas (www.csc.noaa.gov/legislativeatlas): For organizations interested in coastal and marine spatial planning and regional ocean management, the Legislative Atlas provides quick access to the complex set of laws governing the nation's ocean waters. The atlas enables users to pinpoint and view on a map the laws, policies, and jurisdictions that apply to their specific coastal areas. It also offers users the ability to access and download laws in the form of spatial data.

To learn more about the center's ocean planning resources, visit www.cmsp.noaa .gov. ◄

need for new interstate transmission lines.

Offshore wind energy projects exist primarily in Europe, where reduced land availability has pushed developers toward offshore sites. Through 2010 there were approximately 2,950 MW of offshore wind capacity installed in the European Union. Approximately 880 MW were added in 2010 (GWEC 2011). However, high costs for construction, installation, and operations and maintenance have presented barriers to the development of offshore wind in the United States.

Costs are substantially higher as a result of the more complex and material-intensive foundation and support structures required in the offshore environment, significantly greater logistics challenges (work at sea drives up costs both during installation and while performing maintenance), and increased electrical infrastructure cost associated with submarine electrical systems (Junginger et al. 2004; UKERC 2010). The fact that much of the necessary supporting infrastructure, including ports and vessels, has yet to be built in the United States also pushes costs higher, as equipment and vessels may need to be brought in from other parts of the world to do the work. Permitting and regulatory barriers and costs are also significant in the United States. Cost uncertainty resulting from limited to zero experience constructing and operating an offshore wind facility in the United States also forces developers to be very conservative in their cost estimates. This pushes estimated costs even higher, particularly for those projects that are vying to be first.

COMMUNITY WIND

The community wind market segment differs from the others in that it refers specifically to an ownership model rather than a type or size of turbine. As such, community wind projects may

MINWIND, ROCK COUNTY, MINNESOTA

The Minwind community wind projects are prime examples of wind energy projects planned, built, owned, and managed by local residents. Minwind is located in the southwest corner of the state, seven miles from the town of Luverne in Rock County, Minnesota. The area has a long tradition of agricultural production. Minwind started taking shape in 2000, when local farmers began looking for a way to capitalize on the area's wind resources. The community wind concept emerged as a potential strategy for developing this resource in such a way that the majority of the economic benefits would accrue to local farmers and residents.

The group determined it would install two wind energy projects, Minwind I and Minwind II. The two projects were set up as separate limited liability corporations (LLCs); this ownership structure enabled the projects to take advantage of tax credits and other incentives. The local economic benefits of Minwind extended beyond local ownership of the projects. Shares in each project were sold to local investors; according to the LLCs' regulations, 85 percent of shares needed to be owned by local farmers, with the rest available to nonfarming local residents. Additionally, no single person was permitted to buy more than 15 percent of the shares. In only 12 days, all shares were purchased by 66 investors within Minnesota.

Throughout construction, the LLCs used local labor, products, and suppliers whenever possible. "We wanted a farmer-owned project that would bring economic development, get farmers a return on their investment, and use local businesses and contractors to do the work," said Mark Willers, CEO of Minwind and president of Minwind I. Though the Minwind projects were legally organized as LLCs, they are run like cooperatives in that they have voluntary, open membership and democratic member control.

Capital raised from the sale of the shares was used toward development costs of the two projects. Each project also received loans from local banks, and a USDA Farm Bill Section 9006 renewable energy grant covered 10 percent of the installed cost of the turbines (about $180,000). Minwind I and Minwind II then took advantage of a Minnesota renewable production incentive that, over a period of 10 years, pays 1.5 cents per kilowatt hour for wind projects up to 2 MW. (Minwind's turbines are just below this capacity.)

One of the more challenging aspects of this development was negotiating a power purchase agreement with a power purchaser. The LLCs initially had some difficulty finding a power purchaser that was willing and able to work with the community-owned Minwind projects. After months of negotiation, Minwind started a 15-year contract with Alliant Energy, which uses the produced electricity to help satisfy renewable energy standards in the nearby states of Iowa and Wisconsin.

In Minnesota, the state's Public Utilities Commission is responsible for regulating and approving large wind energy facilities of over 5 MW, and small systems under this threshold are permitted by local governments, though they must incorporate commission-prescribed general permit standards in their processes. Eric Hartman, director of Rock County's land management office, explained that the Minwind projects were approved as conditional uses, with a public hearing required for each of the turbines during its development stage. According to Hartman, public opinion grew more and more positive with each subsequent turbine.

The Minwind model for community-owned wind energy has been successful. After the first two projects were built, local interest was at such a high level that advocates began planning for additional Minwind LLCs. Today there are nine Minwind projects, each with an energy capacity of about 1.75 MW. Minwind I and II have two smaller turbines each, while subsequent Minwind projects have just one larger turbine apiece.

To those involved with Minwind, not only is this project environmentally sustainable, it is economically sustainable as well, and it is a profitable business venture for local farmers. According to Willers, the Minwind projects were developed with many objectives in mind, including generating renewable energy, creating local employment opportunities, maintaining group ownership, keeping profits local, and participating in the future. In Minwind's 10 years, it has and continues to accomplish all of these goals.

For more information:

- Minnesota Public Utilities Commission. "Wind Turbine Siting." Available at http://energyfacilities.puc.state.mn.us/wind.html.
- "Minwind: Farmer-Owned." Available at www.windworks.org/articles/minwind.html.
- "Minwind I and II Project Rock County, Minnesota." Available at http://nwcommunityenergy.org/wind/wind-case-studies/minwind.
- Windustry. "Minwind III–IX, Luverne, MN: Community Wind Project." Available at www.windustry.org/minwind-iii-ix-luverne-mn-community-wind-project.
- Windustry Newsletter. 2002. "Minwind I & II: Innovative Farmer-Owned Wind Projects." Available at http://windustry.org/news/windustry-newsletter-fall-2002.

utilize any of the various turbine types (i.e., distributed, midsized, utility-scale, or offshore). The key element of a community wind project is some form of local ownership or equity investment, whether from local residents (e.g., farmers and ranchers), colleges, tribal governments, or local businesses. This sector accounts for about 2 percent of the overall wind industry (Wiser and Bolinger 2011). However, some definitions of the sector include projects owned by municipal utilities and rural electric cooperatives, which boosts the share of the market. In theory, a community wind project may be of any size so long as there is a local ownership component; in practice, community wind projects are often smaller than average utility-scale wind projects.

Community wind projects face many of the same barriers as those in the broader industry; however, due to restrictions on the type of income that can be used to monetize the production and investment tax credits (see page 15), community wind projects have tended to face more financing challenges (Bolinger 2011). In addition, small community wind projects may have difficulty achieving the economies of scale associated with utility-scale projects.

Community wind, however, has been observed to hold some advantages. Notably, community wind projects are believed to enhance social acceptance by encouraging stakeholder buy-in and participation (McLaren 2007) and distributing economic benefits more broadly throughout the economy (Lantz and Tegen 2009).

LOCATIONS

Erica Heller, AICP

As more wind turbine technologies are developed, wind turbines can fit into an increasing variety of settings. Quieter, smaller models can fit into more densely settled areas, while large, tethered turbines have successfully been installed in offshore locations. Different contexts—whether urban, rural, offshore, or off the grid—raise particular issues for wind energy compatibility.

Urban

Small wind energy conversion systems (WECS) can fit in a wide variety of settings, including urbanized communities. San Francisco, Denver, and Chicago are large cities that permit a variety of small WECS. As discussed in Chapter 6, reasonable standards may readily be drafted to address safety concerns and other potential impacts of small WECS that make it possible to site them close to other uses.

A significant challenge for small WECS in urban areas is access to good quality wind. The presence of numerous structures and obstructions in urban areas cause very turbulent wind at many urban sites, limiting the potential of small WECS in these areas. The turbulence and obstructions result in large differences in wind quality over short distances, making site-specific testing very important to cost-effectiveness analysis.

WECS technologies are being developed to take advantage of urban locations and conditions. (See Figure 2.3, page 24.) Some roof-mounted models are designed to be placed in a row along the windward edge of flat commercial and industrial rooftops, where they take advantage of the updraft from the building face. Micromodels may be mounted atop light poles in parking lots. However, wind access and quality remain challenges in such locations. Studies that measure the performance of urban and rooftop wind turbines indicate substantial concerns for many urban sites due to turbulence. Therefore, it is not advisable at this time for local governments to limit allowable small WECS to building-mounted models, even in urban settings.

LINCOLN COUNTY, KANSAS

Lincoln County, Kansas, in the northern central region of the state, is a rural county of 720 square miles with a population just under 3,500. Seeing the economic declines common to many other predominantly agricultural areas of the Great Plains, county officials decided in 2000 to look into their jurisdiction's potential for wind energy. This was before wind energy had gained much national attention, and the state had yet to release wind resource maps and studies, so the county was heading into uncharted territory, recalls Jennifer O'Hare, county attorney: "It was really a grassroots effort. The Economic Development Department and the Board of County Commissioners put up a meteorological tower to explore wind levels in the county, and they invited other counties to join in—but none was interested. It turned out that the wind resources in the county were very impressive, and the county officials approached wind developers." Transmission lines already ran through the county, which further increased its attraction.

The eventual outcome was the Smoky Hills Wind Farm, the state's largest wind energy project, which began operating in 2008. Built in two phases, the project totals 155 turbines with a 250 MW capacity and covers 26,000 acres involving 200 landowners in Lincoln and Ellsworth counties. There is no zoning in the county and no planning or policy documents, so the permitting process consisted mainly of the standard development form required by the Board of County Commissioners and a few additional key negotiations: road maintenance, county land leases, and a PILOT agreement.

The highway department worked out a roads maintenance agreement with the developer that required documentation and repair of any road damage resulting from turbine construction, as well as developer funding of any road upgrades required to truck the turbines to their sites. While residents enjoy the better road conditions, O'Hare points out that the county is now responsible for paying for the long-term maintenance of those roads and has had to adjust its annual budget accordingly.

The board negotiated directly with the developer for lease agreements for the turbines on county land, but since the bulk of the project was built on private land the developer negotiated directly with those landowners. Due to the large amount of open land in the county, the developer was able find enough interested

landowners while avoiding those who were not interested or who opposed the project. Some project landowners host turbines on their property, but the developer also leased additional land as buffers to keep other wind farms from being built nearby.

Finally, the county negotiated a PILOT agreement for the project. The initial agreement was for $275,000 for the year of project completion and $200,000 for the subsequent nine years, with future PILOTs to be negotiated based on project expansion. In Kansas, wind farms are not taxed, but state statutes allow local governments to accept PILOTs instead. O'Hare explains, "PILOTs serve as a 'gift' that offsets some of the project's tax liability and demonstrates the good faith of the developer." But while PILOTs typically go into a community's general treasury, the board of commissioners took an innovative approach, creating the Windpower Economic Benefit (WEB) fund to ensure direct benefits to the entire community. The fund is overseen by a five-member board representing the economic development department, the board of commissioners, and the county's three districts. Each year the interest from the project PILOT is made available as unstructured grants to local groups, including the board itself, the county economic development agency, schools, and community groups. O'Hare reports, "This is the fund's first year; the first application has come in from the hospital, and several other community organizations are preparing applications." The WEB fund bylaws are designed to ensure that community goals will be forwarded through this process. Subsequently, other Kansas counties have followed Lincoln County's lead and used its PILOT agreement as a template.

According to O'Hare, public support for the project is high and was bolstered by the proactive efforts of a local landowner and farmer who advocated for the project early in the development process. Small and rural with no oil, gas, or big business presence, the county benefits from the financial resources that the wind farm brings in and has worked to make sure that they go toward supporting the community's quality of life. O'Hare adds, "For children born in the community after the turbines went up, Kansas has always had turbines—there has never been a time for them when the turbines weren't there. This experience has really opened the eyes of Kansas residents to the renewable energy process and shown them what Kansas has to offer in this regard."

While density and proximity of uses in urban areas can add some challenges for siting and operation of WECS, some of the attributes of urban areas enhance compatibility with WECS. Ambient noise in urbanized areas, such as in industrial districts or near freeways, can mask the additional sounds generated by WECS. Many urbanized areas host an array of visual obstructions such as transmission lines, cell towers, radio antennas, tall buildings, smokestacks, and billboards. It may be easier for urban residents who are used to such visual clutter to accept WECS as part of the skyline than it is for rural residents accustomed to pristine views. Nonresidential urban users with high energy demands may find that installing a WECS is a cost-effective way to reduce energy bills. For example, WECS are a good fit with the character of industrial zones and can provide power to energy-intensive industrial uses.

In urban areas, it may be beneficial or enhance community acceptance to define subcategories of small WECS. Many different sizes of "small" WECS fit different scales of urban uses, from factories to single-family homes. An ordinance that restricts the energy output of small WECS so that it primarily serves on-site needs is one effective way to scale WECS to the use. However, in some instances, a large user, such as a church, might be located in a residential district and use enough energy to justify a WECS large enough to feel out of scale with the neighborhood. Thus, urban communities and those with a great deal of variety in character are more likely to take a more fine-grained approach to scaling size or output of WECS by zoning district.

Height limits in urban areas often need to be higher than in rural areas to ensure that WECS have access to less turbulent wind. In rural areas, it may be possible to simply move WECS horizontally to avoid obstructions and turbulent wind, but in densely developed urban areas with cluttered airspace, the only viable direction to go is up. Mechanisms to protect installed WECS' wind access against future obstruction are also important. Basic structure height limits for each zoning district can establish certainty for potential WECS installers, while exemptions from the district limits allow WECS to clear obstacles. At limited urban sites, such as on areas of high ground, along shorelines, or adjacent to open areas, wind may be quite consistent at lower elevations. In these areas, WECS may not need to be as high to function well.

Rural

Rural locations are the traditional setting for wind turbines. The history of wind turbines begins with windmills in rural agricultural contexts. Today, utility-scale wind farms are located almost exclusively in rural areas where winds are strong and unimpeded. Small wind turbines also continue to be located in agricultural and other rural settings.

Overall, small turbines fit more readily into rural areas than urban locations. Many of the same concerns and possible land-use impacts exist, but the greater separation among rural land uses can help to resolve such issues. Three land-use considerations that merit discussion in small wind energy regulations for rural areas are noise, setbacks, and visual impacts.

Noise impacts in rural areas may be considered more annoying than in urban areas due to lower ambient noise levels. Standards for acceptable sound output typically specify a level 10 to 15 dB greater than the ambient baseline. In rural settings where typical lot sizes are quite large, noise is often measured at the nearest habitable structure rather than at the property line. This measurement location acknowledges that many areas of a large lot may not be regularly used for outdoor occupation or leisure and that turbines have less potential to create nuisance impacts in a cornfield than in a backyard.

Figure 2.3. Small WECS generate enough energy to light this parking lot in Lakewood, Colorado.

Source: Erica Heller

In rural areas with large lots, minimum setbacks are sometimes larger than in urban areas. Larger setbacks can create an additional buffer to reduce concerns about noise and visual impacts and increase acceptance of wind turbines. Where lot sizes are typically larger, most landowners can meet the larger setbacks, and thus it is reasonable to require them. In some rural areas, particularly agricultural ones, setbacks are measured from the nearest habitable structure or public way rather than from the property line, much as with rural noise regulations.

The issue of visual impact of small turbines can be either more or less controversial in rural areas than in more urban settings. In agricultural communities, wind turbines are often readily accepted as a part of the working landscape. Many other rural communities tend to take a minimalist approach to land-use regulation in general, and installation of small wind turbines is accepted along with a wide range of other decisions about the use of private property.

However, unlike urban areas, rural areas can have more pristine visual environments. In some rural communities, small wind turbines are perceived as an unwelcome intrusion in the otherwise wide-open skyline. Small wind turbines may be controversial in these areas due to their potential impact on neighbors' views. In these communities, visual impact analyses may be required or specific viewsheds may be called out for special protection from turbines and other potential visual intrusions. While some communities try to address this concern through restrictive height limits on turbines, this practice severely limits wind turbines' effectiveness. (See Chapter 6.) Height limitations that keep small wind turbines below the level of clean, strong winds are not an effective regulatory solution as they render turbines cost ineffective and thus function as a de facto prohibition.

Off-Grid

Most often, small WECS are grid-connected, meaning they feed into the existing electricity grid rather than directly to the primary use. Wind is variable, seasonal, and may not be well-timed with demand and usage. For example, wind may be best on spring and autumn nights, whereas peak demand occurs on summer weekday afternoons. Where a connection to the grid is available, it is the practical way to capture all the WECS' output, which can be credited against the electricity that is used on-site. In fact, some cities and towns require proof of utility approval of a grid connection as a submittal requirement for a small WECS permit.

However, WECS can also serve areas and uses where a grid connection is not practical or feasible. In some agricultural districts, the popularity of WECS for farming applications stems from the fact that they can provide significant power without the hassle and expense of running long connectors from transmission lines through tilled farmlands. One of the oldest applications of windmills—pumping water—remains useful on agricultural areas because it successfully overcomes the temporal problems of wind energy. Energy generated by WECS is used to pump groundwater into surface ponds whenever the wind blows, for use as needed by gravity feed.

WECS are used in other off-grid applications as well, such as at remote park facilities and residences such as mountain cabins that have no possibility of grid connections. Occasionally, alternative energy "purists" that could connect to the grid prefer not to, as they object to the exchange of electrons with a grid where power is primarily derived from fossil fuels. These users may seek to create off-grid systems.

Where small WECS are not connected to the grid, owners typically address the temporal challenge of wind energy by feeding power into battery systems that store energy during windy times for use later. Often uses that are not grid-connected couple small WECS with other renewable energy options, such as solar, to generate energy in a range of weather conditions and seasons.

UTILITY-SCALE WIND PROJECT COSTS AND ECONOMICS

Kevin Rackstraw

Windpower generation costs have been decreasing over recent decades, enabling a rapid expansion of wind projects worldwide as they have become more competitive with other electricity-generating options. However, both the cost of the average installed wind project (Figure 2.4) and the resulting cost of energy produced (Figure 2.5) have actually risen over the past five years.

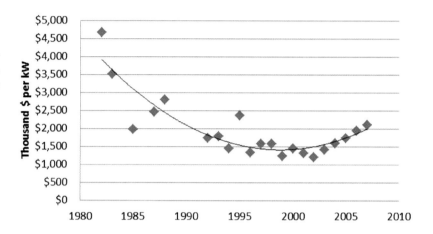

Figure 2.4. Wind power installed cost trends

Source: Wiser and Bolinger 2010

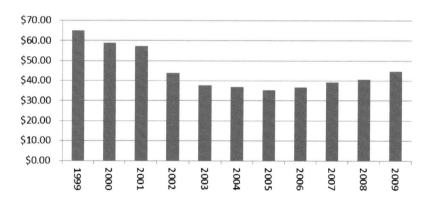

Figure 2.5. Cumulative capacity-weighted average wind power prices

Source: Wiser and Bolinger 2010

Despite these trends, wind power is seen as an attractive option by growing numbers of utilities, as illustrated by its 37 percent annual average growth over the past five years (WWEA 2010, 19). There are a number of reasons for wind energy's continued growth—such as policy, improved turbine performance, the lack of fuel price risk, and many others—but it is also significant that most other electricity-generation options have increased in cost (though natural gas prices have recently declined again).

Basics of Wind Economics

The cost of wind energy is determined by the installed cost of the wind plant; the operating cost of the plant, including the costs of maintaining, repairing, insuring, and financing the project, taxes, and other project-related charges; the set of incentives that affect the final price of the power, the demand for wind power, or both; and the amount of wind that can be captured by the plant.

Installed Cost of the Wind Plant. The main cost elements of a utility-scale wind plant are turbines (generators, nacelles, blades); towers; power transformers, both at each turbine and at a substation; cables for carrying power and electronic signals; substation and switching equipment to allow interconnection into a high-voltage grid; computers, fiber-optic networks, office, storage facilities; construction costs (excavation, foundations, concrete, roads, erection of turbines, trenching of power cables); development costs, including permitting; and financing costs, including construction interest, insurance, legal, and other transactional costs, although the installed cost does not include the cost of permanent financing.

Figure 2.6 shows a breakdown the major cost categories of a typical 100-MW project in 2010.

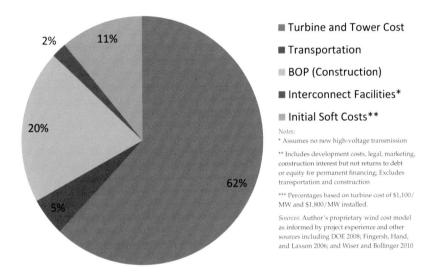

■ Turbine and Tower Cost

■ Transportation

▨ BOP (Construction)

■ Interconnect Facilities*

▨ Initial Soft Costs**

Notes:
* Assumes no new high-voltage transmission

** Includes development costs, legal, marketing, construction interest but not returns to debt or equity for permanent financing; Excludes transportation and construction

*** Percentages based on turbine cost of $1,100/MW and $1,800/MW installed.

Sources: Author's proprietary wind cost model as informed by project experience and other sources including DOE 2008; Fingersh, Hand, and Laxson 2006; and Wiser and Bollinger 2010

*Figure 2.6. Breakdown of installed wind project costs****

There can be substantial variation in the percentages from one project to another or from one year to another. When turbine prices were higher, the percentage spent on turbine hardware was closer to 70 percent. Areas that are relatively easy to construct in (such as flat, open farmland) could see lower construction percentages and, thus, higher percentages in turbine costs. Some projects may also require the construction of new high-voltage transmission infrastructure, which can be expensive depending on voltage and the length of the line. The transmission total could easily be several percent of a project's costs if many miles of new line must be built or if interconnection with transmission lines higher than 230 kilovolts is required. To simplify matters, this analysis assumes that no new transmission will need to be built.

Development Cost, Uncertainty, and Risk. It is important for planners to understand development cost. A developer often has to invest millions of dollars and several years before a project is financed and ready for construction. This investment is very much at risk until the point at which financing is committed and construction has begun. Permitting costs are an important part of that equation, as they can amount to hundreds of thousands if not millions of dollars when local, state, and federal requirements are taken into account. Redundant requirements raise costs and cause delays, so local planners should ensure they are not duplicating state or federal permitting purposes. Clarity and stability of requirements are crucial. From a developer's perspective, very little is more destructive to the development process than uncertain or changing requirements.

Operating Costs. The main operating costs are:

- Maintenance personnel (labor, benefits)
- Maintenance equipment (trucks, tools)
- Spare and replacement parts
- Repair costs for turbines, roads, cables, substation and related electrical equipment, and other project assets
- Insurance
- Management of the operations process, including tracking of plant conditions, and reporting to owners and other relevant parties
- Taxes (income and property taxes principally)
- Land costs (payments to landowners for property usage)
- Other payments to communities and individuals either as compensation or as voluntary community support
- Finance payments (interest and principal on long-term debt, returns to equity, fees)

Other operations costs typically include the cost to deliver energy (transmission) from the project location to a customer, although there are situations where energy is delivered only to the project's point of connection to the grid rather than to a particular customer for a defined price. This sale of energy to the grid (also known as selling to "the market") means a lower cost for the project's owner, since there is no payment for transmission of energy to a distant customer, but it also means that the price of the project's produced energy is subject to energy market variability. Most energy markets change on an hourly or even subhourly basis.

Wind Resource. Many areas of the United States have strong enough winds to generate electricity, but the cost of generating power varies dramatically. The Midwest and Great Plains from the Dakotas down through west Texas have excellent resources, so the projects with the lowest-cost energy generally are in this region, while the east and west coasts have mostly moderate wind resources with a few specific areas that have strong resources (e.g., the Columbia River Gorge in the Northwest or the Tehachapi Mountains in California). Since winds generally are stronger at higher elevations, most projects in the east have been built on top of mountains or ridges. While mountaintop projects are more expensive to construct, energy prices in the eastern United States are also higher, so wind projects can still be competitive there. Wind resources offshore are also excellent on the coasts, though the cost of utilizing that resource is far higher than onshore and has not been competitive to date.

Incentives. Wind energy projects can receive two sources of value from the tax code. The first is five-year accelerated depreciation (Modified Accelerated Cost Recovery System or MACRS), which is also applicable to a wide variety of other high-technology assets such as computers, aircraft, and petroleum drilling equipment. A second tax incentive, which is targeted toward clean energy sources like wind, provides a project owner with a choice of either an investment tax credit or a production tax credit. The current Production Tax Credit (PTC) in 2010 amounted to 2.2 cents/kWh of energy produced and is set to expire at the end of 2012. This tax credit can reduce the cost of the resulting output by 20 to 30 percent.

The PTC is designed to be extremely difficult for individuals to use, so typically only large corporations have the right kind and quantity of tax liability to utilize it efficiently. For recent projects (2009–2011), a cash grant has been available in lieu of the tax credits. This was created in response to the economic recession and a lack of tax liability by corporations and is not likely to be extended after it expires at the end of 2011. The cash grant opened ownership to many new entities besides large corporations, which introduced new financial models and brought many new (and smaller) players into the business. The cash grant/investment tax credit simplified the calculation of energy cost since it could just be deducted from the cost of the project once it was complete. The PTC generates tax credits over a 10-year period, but since the user of it might not have a tax liability every year, the final cost of energy produced is uncertain.

States also provide incentives for wind projects, often in the form of a corporate income-tax credit, but some provide property tax abatements or exemptions, sales tax exemptions, or performance incentives much like the federal PTC. An excellent summary of the various state incentives—as well as details about them and their enabling legislation—can be found at www.dsireusa.org/summarytables/finre.cfm.

Other incentives include state or federal requirements to purchase wind energy. These mandates are often called Renewable Portfolio Standards (RPS) and are in place in 30 states, while another seven states have voluntary standards in place. (See www.dsireuse.org/summarytables/finre .cfm.) The federal government also aims to purchase a portion of its energy from green sources—5 percent in 2011, going up to 7.5 percent in 2013. The state RPS policies are generally considered primary drivers of renewable energy capacity in the United States, although some states, such as Texas, have already exceeded their requirements and have added new capacity (mainly wind) for economic reasons. (Electricity prices in Texas are set at the margin by natural gas, which meant several years ago that electricity prices were quite high. In this environment, wind energy was a very attractive option.)

Wind Cost Trends

The reason for wind energy's cost increases since 2002 range from a declining exchange rate to increased commodity prices for steel and copper to increased margins extracted throughout the wind turbine supply chain, including from key suppliers of components such as gearboxes and from experienced construction firms that benefited from the extremely hot market from 2007 to 2009. Going forward, the U.S. market may be somewhat more insulated from exchange rate fluctuations because more blades, gearboxes, bearings, generators, electronics, and towers are being made domestically. Historically, European firms have dominated U.S. turbine supply, with only a couple of significant American turbine suppliers in the field. Today, that picture is quite different (Figure 2.7).

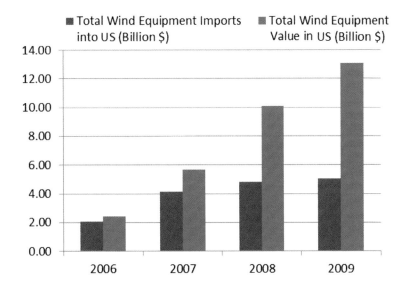

Figure 2.7. Imports as a percentage of total wind equipment costs

Source: Wiser and Bollinger 2010

Impacts of Economies of Scale on Wind Costs

Figure 2.4 (page 26) shows declining wind power prices from 1999 through about 2005, while Figure 2.5 (page 26) shows installed project costs rising in 2001 and 2002. That wind power prices fell despite rising installed costs is principally due to economies of scale, both in terms of project size and in terms of turbine size. Another factor is the maturation of the industry, including the entry of larger, better-capitalized developers who were able to improve efficiencies in the development process as well as in the financing and construction of projects. From the practicing planner's point of view, this material should be useful in providing context for judging claims by developers that larger projects are necessary to optimize the project's economics.

Turbine Size. The most dramatic economy of scale achieved by the wind industry is the increasing size of wind turbines, mainly in terms of the height of the towers, size of the generators, and length of the blades. Essentially, the latest turbines are able to squeeze more energy out of any given tower and foundation (Figure 1.1, page 2). Longer blades, for instance, increase the area from which energy is captured by the turbine. Think of the blade as the radius of a circle that is "harvested" of energy by the moving blades. For every increment of additional blade length (r), the circle increases by a squared factor (a circle's area is defined as πr^2). In some cases, the increased energy capture of the blades will also allow a larger generator and gearbox to be used so that the production capacity of any given turbine location can be much higher with only a moderate increase in costs, all of which leads to an overall lower cost of energy.

Turbine Height. Size also enables better economics because wind speeds tend to increase with distance from the earth's surface, a factor that is called wind "shear." There are limits on height, as towers get exponentially more expensive as they go up. Still, the average height of turbine towers has grown over the last decade from about 65 meters to about 80 meters (213 to 262 feet). Increasingly, towers are being built up to about 100 meters (328 feet) to try to improve energy production, and thus improve project economics, but cost is not the only limitation on tower height, as planners know. There are viewshed, aviation, radar, and wildlife issues that also serve to limit tower height. Still, as technologies improve (and as towers and turbines get lighter so that foundations do not have to be as strong), there will be economic pressure to increase the height of towers to take advantage of wind shear.

Larger Project Size. Economies of scale can be achieved from larger projects, on account of the larger capacity of each individual turbine and the number of turbines built at any given location. It is very expensive to bring in the big cranes that are needed to erect wind turbines, as well as to mobilize crews and buy turbines and related material, so that the larger the project, the more efficient it is.

Understanding the Energy Price Context

To understand wind energy's continuing competitiveness with other energy technologies, we have to look a little more closely at the numbers and at the broader context. "Competitive" does not necessarily mean that wind energy provides the lowest cost of electricity at a given moment. When utilities or other major buyers of energy look at costs, they are looking at them over some defined period—usually at least a few years, sometimes as many as 20. For instance, when natural gas prices are low, gas-fired projects can claim extremely low costs. However, natural gas historically has been an extremely volatile commodity (Figure 2.8), so over the life of a natural

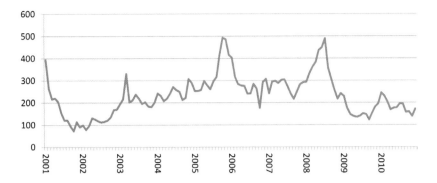

Figure 2.8. Gas price index, 2001–2010

Source: Bureau of Labor Statistics (http://data.bls.gov)

gas–fired plant, the cost might be quite a bit higher much of the time.

This volatility requires some party to take the risk that natural gas prices might spike, and energy buyers are often hesitant to do so. For that reason, it is not easy to buy natural gas for long periods of time into the future—most contracts last only up to a few years—because sellers demand a substantial premium for that risk. The typical utility customer (ratepayer) who is a home owner (or renter) takes that risk, as utilities always try to pass that cost on to the ratepayer.

The cost of other conventional sources, such as coal, rose enough over the

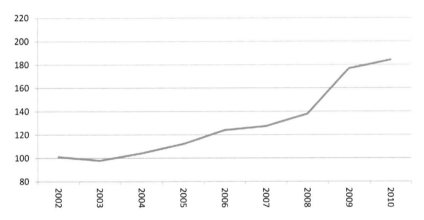

Figure 2.9. Coal price index, 2002–2010

Source: Bureau of Labor Statistics (http://data.bls.gov)

last 10 years that wind's relative position remained strong (Figure 2.9).

Coal also has some price volatility associated with it, particularly eastern coal from the Appalachians. (See www.ferc.gov/market-oversight/mkt-electric/overview/elec-ovr-coal-bs-prb-pr.pdf.) One of the biggest uncertainties for power sector investment today is the potential implementation of some kind of carbon control, whether a cap or a tax. Opinions on the likelihood of carbon constraints, as well as the cost impacts of any such controls, vary widely, but the prospect does introduce some additional uncertainty that investors and utilities have to take into account.

Wind energy, by contrast, has no fuel cost and has relatively low operating costs, so once a project is built (and barring major unanticipated equipment failure), the owner can have a high degree of confidence in the cost of energy for many years into the future. Since wind energy's costs are primarily hardware costs that are heavily influenced by the prices of steel, copper, and other commodities, they vary up to the point of turbine purchase and delivery, but wind technology typically has limited exposure to such costs over the project's operating life. Like other technologies, wind components are subject to failures due to unplanned stresses, manufacturing issues, and wear and tear. Most projects will build a certain amount of unexpected component failures into their project pro formas. Failures that are not covered through third-party insurance, warranties, or self-insurance do expose the project owner to commodity and other market pricing risks at the time of the failure. This exposure can usually be mitigated through insurance or extended warranty products but in most cases cannot be eliminated over the full life of the project.

One of the best ways to think about wind energy's value to electricity buyers and users (that is, its competitiveness) is as a portfolio diversification tool. Much like most investors want price-stable products (e.g., certain bonds, certificates of deposits, cash, etc.) in their portfolios to offset the volatility of stocks, wind energy's fixed-price character fits well in an energy portfolio with more volatile assets. At times of high natural-gas prices, wind energy is often selected by utilities and other electricity buyers as the most economic option. At times of low natural-gas prices, wind energy provides the hedge value that stable-priced investments do in a portfolio.

The general rule has been that wind energy will be most competitive with natural gas at five or six dollars per million BTU (mmbtu) or higher, though this does depend on the price of turbines. Natural-gas prices are approximately $3.50 per mmbtu in the third quarter of 2011, and so natural gas has a near-term price advantage. However, given natural gas's price volatility, wind energy can still be a highly valued asset in a buyer's portfolio.

Cost Comparison of Electricity Options

Those interested in knowing more about wind energy's competitiveness often ask for comparisons of the various major generating options. Unfortunately, comparative energy costs are notoriously hard to do fairly; many studies exist, but they use very different methodologies. It gets even more complicated when studies add costs for "firming up" wind's variability from changes in wind speeds. The cost of such firming can vary from one region to the next, but in most areas such intermittency costs are relatively minor, on the order of one-tenth of a cent per kWh to half a cent per kWh. (See www.awea.org/learnabout/publications/upload/Reliability-Factsheet-March-2011.pdf and Smith et al. 2007.) Some recent studies can be found at www.sourcewatch.org/index.php?title=Comparative_electrical_generation_costs. However, most such studies show wind as highly competitive with other sources over a 10- or 20-year period, particularly when emissions, hazardous waste, and fuel price volatility are taken into account. (See pro forma on page 33.)

The following pro forma provides an example of expected revenues and expenses for a 50 MW wind energy project over a 20-year period. Technology and financial assumptions are included as well as expected incentives, helping to explain the year-to year variation in return on investment.

Technology Assumptions

Project Capacity (MW)	50
Capitalized Cost ($/kW)	1,978
Fixed O&M ($/kW)	39
Fixed O&M Escalation	1.5%
Variable O&M ($/MWh)	6
Variable O&M Escalation	1.5%
Capacity Factor	35%

Financial/Economic Assumptions

Debt Percentage	60%
Debt Rate	7.5%
Debt Term (years)	10
Economic Life (years)	20
Depreciation Term (years)	5
Percent Depreciated	85%
Cost of Generation Escalation	0%
Tax Rate	35%
Cost of Equity	9.5%

Incentives

PTC ($/MWh)	22.00
PTC Escalation	1.5%
PTC Term (years)	10
ITC	-

Levelized Cost of Generation: 68.26

Year	1	2	3	4	5	6	7	8	9	10
Annual Generation (MWh)	153,300	153,300	153,300	153,300	153,300	153,300	153,300	153,300	153,300	153,300
Revenues per MWh (incl RECs)	68.26	68.26	68.26	68.26	68.26	68.26	68.26	68.26	68.26	68.26
Operating Revenues	10,463,510	10,463,510	10,463,510	10,463,510	10,463,510	10,463,510	10,463,510	10,463,510	10,463,510	10,463,510
Fixed O&M	1,938,774	1,967,856	1,997,374	2,027,334	2,057,744	2,088,610	2,119,939	2,151,739	2,184,015	2,216,775
Variable O&M	919,800	933,597	947,601	961,815	976,242	990,886	1,005,749	1,020,835	1,036,148	1,051,690
Operating Expenses	2,858,574	2,901,453	2,944,974	2,989,149	3,033,986	3,079,496	3,125,689	3,172,574	3,220,162	3,268,465
Interest Payment	4,450,562	4,135,970	3,797,783	3,434,233	3,043,416	2,623,288	2,171,650	1,686,140	1,164,216	603,148
Principal Payment	4,194,561	4,509,153	4,847,340	5,210,891	5,601,707	6,021,835	6,473,473	6,958,983	7,480,907	8,041,975
Debt Service	8,645,123	8,645,123	8,645,123	8,645,123	8,645,123	8,645,123	8,645,123	8,645,123	8,645,123	8,645,123
Tax Depreciation (5-yr MACRS)	16,813,234	26,901,175	16,140,705	9,684,423	9,684,423	4,842,211				
Taxable Income	(13,658,860)	(23,475,087)	(12,419,952)	(5,644,294)	(5,298,315)	(81,485)	5,166,172	5,604,797	6,079,132	6,591,897
PTC	3,372,600	3,372,600	3,525,900	3,525,900	3,525,900	3,679,200	3,679,200	3,679,200	3,832,500	3,832,500
ITC	-	-				-			-	
Taxes	(8,153,201)	(11,588,880)	(7,872,883)	(5,501,403)	(5,380,310)	(3,707,720)	(1,871,040)	(1,717,521)	(1,704,804)	(1,525,336)
Total	7,113,014	10,505,815	6,746,296	4,330,641	4,164,711	2,446,611	563,738	363,334	303,028	75,258

Year	11	12	13	14	15	16	17	18	19	20
Annual Generation (MWh)	153,300	153,300	153,300	153,300	153,300	153,300	153,300	153,300	153,300	153,300
Revenues per MWh (incl RECs)	68.26	68.26	68.26	68.26	68.26	68.26	68.26	68.26	68.26	68.26
Operating Revenues	10,463,510	10,463,510	10,463,510	10,463,510	10,463,510	10,463,510	10,463,510	10,463,510	10,463,510	10,463,510
Fixed O&M	2,250,026	2,283,777	2,318,034	2,352,804	2,388,096	2,423,918	2,460,276	2,497,180	2,534,638	2,572,658
Variable O&M	1,067,465	1,083,477	1,099,730	1,116,226	1,132,969	1,149,963	1,167,213	1,184,721	1,202,492	1,220,529
Operating Expenses	3,317,492	3,367,254	3,417,763	3,469,030	3,521,065	3,573,881	3,627,489	3,681,902	3,737,130	3,793,187
Interest Payment										
Principal Payment										
Debt Service										
Tax Depreciation (5-yr MACRS)										
Taxable Income	7,146,019	7,096,256	7,045,747	6,994,481	6,942,446	6,889,630	6,836,021	6,781,609	6,726,380	6,670,324
PTC										
ITC										
Taxes	2,501,107	2,483,690	2,466,012	2,448,068	2,429,856	2,411,370	2,392,607	2,373,563	2,354,233	2,334,613
Total	4,644,912	4,612,567	4,579,736	4,546,413	4,512,590	4,478,259	4,443,414	4,408,046	4,372,147	4,335,710

Source: Rackstraw Consulting, LLC

CHAPTER 3

Addressing Concerns

Eric Lantz with Charles Newcomb

 Surveys of the general public typically find broad support for wind energy; however, siting specific wind energy projects remains a challenge (Huber and Horbaty 2010). Local opposition to wind energy projects and subsequent siting challenges persist for many reasons. Wind energy is a low-density resource, which increases its visibility across the landscape and results in far more individual siting decisions than a conventional highly centralized power-generating station. Moreover, wind power introduces a moving element into the landscape (Wustenhagen et al. 2007). In rural areas, where wind resources tend to be best, a wind energy installation may be the first or most visible industrial development in a given locality. Individual community responses to these and other issues may be influenced by factors including demographics, culture and history, the local economy, or interactions and experience with wind industry representatives (Huber and Horbaty 2010).

Common wind energy concerns tend to fall into two major categories: impacts on quality of life, including potential degradation of the landscape and nuisance-related concerns, and impacts on the environment, primarily ecological and wildlife impacts. Within these broad categories are numerous specific concerns ranging from the distribution of economic benefits (e.g., landowner lease payments) to the potential for decreased property values and avian and bat fatalities.

The wind energy industry tends to label people who oppose its projects as "NIMBYs" or individuals who have taken "Not in My Backyard" positions. However, the literature suggests that this phenomenon rarely exists; rather, people typically have real concerns that if addressed properly can be resolved (e.g., Jones and Eiser 2009, Wolsink 2006, Devine-Wright 2005). In this context, siting and planning for wind energy development requires sensitivity to the concerns of a diverse array of stakeholders. It also requires attention to process as well as to substantive issues (Huber and Horbaty 2010).

This chapter discusses specific wind energy issues that tend to concern individuals and communities, as well as typical mitigation strategies.

ENVIRONMENTAL AND ECOLOGICAL CONCERNS

Environmental concerns about wind development have historically focused on direct avian impacts, such as raptor-turbine collisions. More recently, these concerns have broadened to include other wildlife, including bats, prairie chickens, and sage grouse. Questions about ecological fragmentation, land requirements, and the magnitude of emissions reductions resulting from deployment of wind power have also been raised. To some extent, wildlife impacts fall under traditional state and federal regulatory schemes; however, issues such as habitat fragmentation may not generate attention from regulators unless a specific species is listed as endangered or otherwise protected. In general, concerns lie primarily at the species or population level, but for endangered or migrating species impacts to individual animals are also a concern.

Source: Kern County, California

Wildlife

Direct impacts on avian and bat populations continue to be a significant concern for the wind energy industry and regulators. Fatalities resulting from direct collisions are important, but displacement and habitat fragmentation—caused when animals avoid wind farms—have also emerged as issues, particularly for species that might be candidates for listing as endangered. In this regard, projects proposed in grasslands have caused the greatest concern, due to the potential for significant avoidance of projects by prairie chickens and sage grouse (Shaffer and Johnson 2008). There is concern that these species may be uniquely sensitive to infrastructure installations and could avoid nesting or brooding near them (NWCC 2010).

A great deal of biological data has been collected on the impacts of wind facilities on avian populations. Literature surveys have estimated bird fatalities to range from 0.95 to 14 per MW per year (NRC 2007; NWCC 2010). Data on raptor fatalities clearly show less gross impact, with roughly 0 to 1 fatality per MW per year; however, these events are often higher profile, as raptors tend to garner more public interest and in some cases are protected (e.g., bald and golden eagles). Data collection on bat fatalities has taken place only in the last few years, following large numbers of fatalities at wind facilities in the eastern United States. A review of more than 40 bat-related studies suggests wide-ranging fatality rates, from 0 to 40 per MW per year (NWCC 2010). Early studies (e.g., Arnett et al. 2008) indicate that bat fatalities are highest among migratory species during migration periods in late summer and fall. Additional data are needed on bat fatalities in other parts of the country and among various species of bats.

The wind industry often seeks to put these deaths in context by comparing the numbers of fatalities resulting from wind turbines to those resulting from other human activities. For example, the fraction of bird fatalities from wind turbines has been observed to be several orders of magnitude below those from vehicles, windows, communication towers, pollution, house cats, and other anthropogenic causes (Erickson et al. 2005; NRC 2007). Such perspective is important, but the relative impact of wind energy could grow as the industry looks to install tens or hundreds of additional gigawatts of capacity.

As the industry has acquired greater knowledge of its impacts on avian populations and developed mitigation strategies, concern over these impacts has generally diminished among biologists. Nevertheless, the emergence of impacts to bats and potential impacts to the sage grouse and prairie chicken is expected to support continued research of industry impacts on wildlife. Moreover, threats to listed endangered species or protected species such as bald and golden eagles must be seriously considered and planned for, as legal challenges on the basis of impacts to protected species can be an effective means of stopping projects.

Mitigating Wildlife Concerns. Understanding the biological context for new wind developments is critical to reducing risk to avian, bat, and other wildlife populations. To a large extent, improved project planning and siting practices, based on more than two decades of collision research, has facilitated dramatic reductions in the rates of avian fatalities at most wind plants. In addition, preliminary research focused on altering wind plant operations during periods of low wind speed and at specific times of the day or year has shown promise, reducing bat fatalities by as much as 80 percent (Arnett et al. 2009; Baerwald et al. 2009). However, additional research is necessary to confirm these rates of effectiveness. Future research is expected to focus increasingly on interactions between animals and turbines to evaluate how wind projects affect normal wildlife behaviors. Some developers have also begun to compensate for their impacts by developing mitigation banks—typically some form of habitat restoration area or preserve designed to offset

WASHINGTON COUNTY, MARYLAND

Washington County, Maryland, has made a concerted effort to "go green" for its residents. Officials, with significant public input, have tailored the county zoning ordinance to accommodate alternative energy development in response to local demands. The result is a planning environment where alternative energy development is welcomed at a scale comfortable for both constituents and the development community.

Washington County is located on the eastern end of the Maryland panhandle, bordered by Pennsylvania to the north and West Virginia to the south. Noticeable interest in wind energy arose there around 2007, coinciding with favorable state and federal tax incentives schemes, as well as the maturation of the wind energy industry. Two brothers, both farmers from Smithburg in the eastern part of the county, were the most vocal advocates. At that time, Washington County's zoning ordinance defined windmills in the context of dairy farm water-pumping machines, and county officials realized they would need to update their ordinance to include wind turbines for electricity generation.

Frostburg State University, near Cumberland, developed a program to evaluate wind and solar energy resources in Washington County. It compared the efficiency of wind to solar generation and found that most of the county was not a significant source of wind energy; only two areas (ridges on the eastern and western sides of the county) had an average wind speed of at least 14 mph. However, because rapid advances in wind turbine technology have made power generation more attractive in areas with less wind, county officials decided to allow both solar energy and small wind energy through a text amendment to the existing zoning ordinance, while restricting commercial wind farms such as those found in neighboring counties.

On June 16, 2009, Washington County successfully amended its zoning ordinance to permit small wind energy and solar collection systems as accessory uses in all districts. The ordinance established wind energy systems setbacks equal to turbine height plus 20 feet from rights-of-way and property lines, and it set limits of two turbines per parcel, though in agricultural and conservation districts the limit was set at twice the amount of the property's annual electricity used.

Public involvement was largely positive. Energy consultants convened meetings to raise awareness of alternative energy, which attracted a core group of interested citizens who provided input to county officials during the process. Officials tried to address every concern brought forward. Sound was the biggest issue for some, though it was not as important to people in rural areas. This proactive work meant few people raised issues at the time of ordinance adoption.

More recently, the City of Hagerstown adopted new ordinance language for both solar energy and wind energy development. Because Hagerstown is more urban than most of unincorporated Washington County, ground-mounted wind turbines are restricted to commercial/industrial or civic uses, while residential areas may have one "microwind" system per building limited to 15 kW of energy output and 10 feet in height above the highest point of the building.

Nearly two years after the approval of Washington County's alternative energy provisions, officials report that solar energy has proven to be much more popular than wind energy. In fact, only two residential wind turbines have been constructed (by the two farmer brothers from Smithburg), while installation of solar energy systems has been much more widespread, particularly among big-box retail stores including Staples, Kohl's, and a locally owned copy store. Even with a county residential stimulus program in place the first six months after the ordinance text amendment was approved, no additional wind turbines were built. Meanwhile, only solar projects have taken advantage of a 100-percent permitting fee credit for alternative energy development in the county.

County officials maintain that, despite the paucity of wind energy development in these first two years, installers must be more strongly regulated. They are concerned that installers will try to convince people that wind energy development is a viable option in areas where the data do not indicate adequate wind resources, resulting in abandoned residential wind turbines marring the rural landscape. Officials have considered amending the ordinance to hold property owners responsible for either reactivating or removing abandoned wind turbines, but no action has yet been taken. Washington County demonstrates the tensions inherent in areas with local interest in alternative energy sources yet little wind resource.

For more information:

- Hagerstown Land Management Code (2011). Article 4, Section (K)(8), Supplementary Regulations—Alternative Energy Sources/Generators. Available at www.hagerstownmd.org/Assets/Plan_Dev/Article-4_JthruP.pdf.

- Washington County Zoning Ordinance (2010). Article 4, Section 4.24, Small Wind Energy Systems. Section 4.25, Solar Collection Systems. Available at http://washco-md.net/washco_2/pdf_files/legal/ZoningOrd_Rev16.pdf.

the impacts of their projects. Including provisions for mitigation banks in planning efforts may assist in resolving objections to wind projects put forth on the basis of wildlife impacts.

Land Requirements

A modern wind energy facility consisting of hundreds of megawatts of power generation will cover large swaths of land. As a rule, 5 MW of wind power can be sited per square kilometer. With a modern 2–3 MW turbine, this translates to between 70 and 130 acres per turbine. Total land use for individual projects, however, is ultimately determined by turbine spacing requirements and local siting constraints. Turbine arrays are designed to minimize production losses due to wind turbine wakes. Terrain and the prevailing wind direction are critical features in determining layouts that minimize production losses. At the same time, project developers must take into account property lines and lease agreements with landowners, setback requirements, and other landscape features, including roads. In some cases, local siting constraints are the primary determinants of the minimum land area requirements. Analysis of actual wind projects has shown individual project density to vary from 1.0 to 11.2 MW/km², with an overall average of 3.0 ± 1.7 MW/km² (Denholm et al. 2009). Despite the relatively large total footprint of a wind energy facility, the actual land requirements for wind turbines and associated infrastructure (e.g., access roads, operations and maintenance facilities, substation) are rather modest. Only about 2 to 5 percent of the total land footprint of the facility is typically removed from service; the remaining land area may be used for its traditional purpose(s), including farming and ranching.

Source: Kern County, California

Mitigating Land Requirements Concerns. Turbine spacing requirements suggest that the total land area required for wind energy facilities is unlikely to change significantly in the future. However, the industry has reduced its footprint by upscaling wind turbine technology. Individual turbine power-generating capacity has grown from tens of kilowatts to multi-megawatts. Continued upscaling is likely to maintain these trends, further reducing the number of turbines necessary to achieve a specific project power-generating capacity in a given area. Providing maps that show the scale of wind energy projects relative to the total available land, as well as the number of turbines expected to be placed within a given area, can be an effective strategy to illustrate that even widespread deployment of wind energy will not result in total coverage of the landscape.

Emissions

Advocates of wind power often highlight the greenhouse gas and other emissions benefits of wind energy. Critics often raise questions about life-cycle and system-level emissions impacts from adding wind energy to the power grid. Such questions are focused more on the overall justification for wind power than on planning per se, but understanding them is important to developing a broader understanding of wind power.

Power production from wind turbines generates no emissions. However, wind energy projects do generate emissions at different points in their life cycle. Detailed life-cycle assessment looks to capture all embodied emissions associated with wind energy facilities. A complete analysis needs to take account of the emissions associated with extracting and producing raw materials (e.g., steel), manufacturing and transporting materials, and installing, operating, and eventually decommissioning a wind energy facility. Such analyses generally suggest that emissions from wind energy are relatively low, on the order of 4.6 to 27 grams of CO_2 equivalent (gCO_2eq) per kWh (Vestas 2006; Voorspools 2000). This is in line with other renewable energy technologies, which have a median value of 4 to 46 gCO_2eq/kWh. By contrast, the median values for fossil fuels range from 469 to 1001 gCO_2eq/kWh (Edenhofer et al. 2011).

Wind power plants also affect the power system in which they operate. Analysis of system-level emissions impacts focus on how introducing wind energy, which can have variable output levels, affects total system emissions, including changes in emissions rates for conventional generation. It has been claimed that the variability of wind energy leads to decreased operational efficiency among the existing generation facilities, thus increasing their emissions. In fact, a detailed literature review has found that the

LOCUST RIDGE, PENNSYLVANIA

The Locust Ridge wind energy project, developed by Iberdrola Renewables, is spread over 8,000 leased acres in Schuylkill County in Mahanoy, West Mahanoy, and Union townships, as well as Conyngham Township in Columbia County, Pennsylvania. It was built in two phases: Phase 1, begun in 2006, consists of 13 turbines of 26 MW total capacity; Phase 2 added 51 turbines and 102 MW of capacity. Locust Ridge is the only wind energy facility in Mahanoy Township and the only completed wind farm in Schuylkill County, though a new county project is currently in development and another has been proposed. Phase 1 of Locust Ridge was developed by Joseph Green, now a project manager for Iberdrola Renewables and a local resident who knew the wind energy potential of its steep slopes and ridges firsthand. "I'm from this area and live here, so I knew the landowners and the land and the local context. I'd hunted on the land, so I knew that the wind was there and the transmission lines were there, that it was a large tract of land without competing uses and that there were unlikely to be conflicts with environmental issues in this area," he says.

Sharon Chiao, chair of the Mahanoy Township Board of Supervisors, explains that the township saw the wind farm project as a promising exemplar of the future of clean energy production, something especially attractive in a region scarred by more than 200 years of coal mining. However, at that time the project was proposed, the township did not have its own zoning ordinance, and its subdivision and land development ordinance (SALDO) never contemplated wind energy development, according to Michael Peleschak, project manager at Alfred Benesch and Company, which acts as the municipal engineer for the Township. "The biggest challenge," he says, "was that there were no rules in place for this type of development. The township saw the project as a benefit, so we were trying to move forward on the project yet stay within the ordinance and make sure that regulatory safeguards were in place."

The process of negotiating development approvals in this uncertain territory was complex. Because Mahanoy Township did not have a zoning ordinance, Green first needed to obtain a county zoning permit and bring it to the Township to obtain a land development permit. However, the county's comprehensive plan did include a section encouraging development of wind energy projects. Green says, "For Locust Ridge I, we had to get a variance through the standard variance proceedings. We had to demonstrate that we had the right to get the variance by showing officials that a wind farm was reasonably similar to other permitted uses. We looked at what was currently permitted: cell towers, utilities, and smokestacks were all vaguely similar to wind farms. Then we had to go beyond that to prove that this was not a hardship we'd imposed on ourselves and that there were no other competing uses for the property."

Once the county zoning permit was obtained, the county and township SALDOs provided checklists of items necessary to obtain the land development permit. Green also had to obtain various state and federal approvals, including a grid interconnection agreement from the regional transmission administrative body; Federal Aviation Administration permits for turbine lighting and signaling; National Pollutant Discharge Elimination System (NPDES) permits addressing state environmental concerns; and Pennsylvania Department of Transportation permits for state highway access. According to Lisa Mahall, county engineer and real estate director for Schuylkill County, once the zoning permit was obtained the county's role was largely to review the project's land development and stormwater plans and make sure that the developer had met all requirements and obtained all the permits; the bulk of the regulatory approval process took place at the township level.

Peleschak describes the township land development permit as a three-plan process—sketch, preliminary, and final—with each stage moving the project forward and delving deeper into specifics. The sketch plan lays out the general idea for the project and helps the municipality identify any potential conflicts or impacts, while the preliminary plan develops the plan design in more detail and resolves any identified conflicts. Green remembers that the unfamiliarity of this land use required extensive meetings with county and township officials and staff. James J. Rhoades Jr., project manager for Alfred Benesch, emphasizes the importance of communication: "The best part about the project was that the developer was very cooperative and upfront with us—he was very proactive about addressing any issues, and his engineer was proactive as well. He really wanted the project to be successful, so he came and asked us how to make it work."

Source: Iberdrola Renewables

The main concerns for both the county and township were stormwater management and safety. Five turbines were sited in the watershed for the town's public water supply, and the project was close to several drinking-water reservoirs, so it was vital to prevent any oil or chemical spills during construction as well as to control erosion and sedimentation. Green used infiltration measures rather than pipes to manage stormwater runoff, which require less infrastructure construction and are easier to maintain. The Township required the substation areas to be well protected with fencing, as the town wanted to make sure future hunting use of the ridge-top areas could continue. Peleschak remembers careful negotiations on turbine siting: "The turbines needed to be set back from the roadways for safety, in case something

(continued on page 42)

(continued from page 41)

happens and a turbine comes crashing down." At the time, Pennsylvania did not have a state-level building code, so the town made its building permit contingent on plan submission and fee payment.

Since that time, both Schuylkill County and Mahanoy Township have adopted ordinances addressing wind energy facilities both large and small. Schuylkill County's zoning ordinance permits small accessory wind turbines for on-site residential use in all districts subject to basic standards providing for setbacks from lot lines and street rights-of-way equal to the height of the turbine; noise and height restrictions; and abandonment. Renewable energy facilities, or wind turbines as a principal use, are permitted by right in industrial and mining districts and as special exception uses in other districts, including agricultural. In 2007, Mahanoy Township adopted a stand-alone Wind Energy Facility Ordinance that designates wind energy facilities as permitted uses in commercial, industrial, and agricultural districts and as special exceptions in residential and conservation districts. The ordinance sets out application requirements for the required wind energy permit and provides standards addressing design and installation, setbacks and waiver options, use of public roads, noise and shadow flicker and waiver options, decommissioning, and public communications. Though the township has not yet been approached for another large wind energy project, it is now prepared to address this use in future. Chiao also notes that in the future the township will be more aggressive about pursuing remuneration from wind energy developers, which pay no taxes on the power their turbines produce. For the Locust Ridge project, the Township charged a registration fee of $1,500 per turbine, which Chiao now believes was too little. "Developers have to understand that they ask a lot of communities, and on these multimillion-dollar projects they should give more back to the municipality."

Both the Benesch engineers and Green agree that having a good wind energy ordinance in place is vital to the success of wind energy implementation. Rhoades encourages municipalities to adopt wind energy provisions first, not to wait until they are approached with a project, saying, "It's important to have rules already in effect so the developer knows what to do before they walk in the door." Green emphasizes that approving wind energy facilities does not have to be a difficult process: "There are existing fair and balanced ordinances out there that take into consideration all stakeholders but aren't weighted against development." He recommends Pennsylvania's model wind farm ordinance as a good example and suggests, "If you want to encourage wind energy, adopt standards that protect your community and clearly lay out the requirements for developers. This removes the regulatory risk; where the rules are spelled out, developers know that if they follow the rules there is a reasonable certainty that the project will move forward." Green adds, "Good developers are accustomed to following the rules—setbacks, noise regulations—to avoid any problems. Regulations do add costs, but the results are good for everyone."

For more information:

- Pennsylvania Department of Environmental Protection, Office of Energy and Technology. "Model Wind Ordinance for Local Governments." Available at www.depweb .state.pa.us/portal/server.pt/community/wind/10408.

- Schuylkill County Zoning Ordinance (2010). Article 4, Additional Requirements for Specific Uses. Section 402(56), Additional Requirements for Specific Principal Uses – Wind Turbines, Other than is Allowed for Wind Turbine as an Accessory Use by Section 403. Section 403(D)(12), Additional Requirements for Accessory Uses – Special Standards – Wind Turbines. Available at www .co.schuylkill.pa.us/Offices/PlanningZoning/PlanningZoning.asp. ◀

reduction in theoretical maximum fuel savings for power systems in which wind provides 20 percent of the total electricity generation has been estimated to be on the order of 0 to 7 percent (Gross et al. 2006). At the moderate penetration levels in place today, efficiency losses throughout the power system resulting from wind energy's variable output only marginally offset the broader emissions savings from wind energy (Gross et al. 2006). (Since emissions are a function of fuel consumption and changes in efficiency directly drive fuel consumption, a modest reduction in system efficiency is equivalent to a modest increase in systemwide emissions.)

Mitigating Emissions Concerns. There may always be some emissions associated with raw material extraction, manufacturing, installing, maintaining, and decommissioning wind energy facilities. However, increasing the efficiency of each process and relying on increased percentages of low-emissions energy to power these processes are likely to reduce life-cycle emissions over the long term. In addition, as power systems are able to more efficiently handle variable output generation, system-wide emissions benefits are also likely to increase. In the end, emissions concerns may be best addressed by providing accurate and reliable reports detailing the actual emissions reduction value of wind energy.

QUALITY OF LIFE CONCERNS

Human and social impacts are primary sources of concern among communities and localities considering wind energy projects. Of greatest concern are often those impacts that affect the ability of inhabitants to feel comfortable in and around their homes. Such concerns often focus on changes to local landscapes and aesthetics, changes in ambient or background sound levels, creation of shadow flicker, and nighttime obstruction lighting. Residents are often also concerned about how these changes might affect individual property values.

At the same time, not all wind energy impacts are negative. Lease payments are typically paid out on an annual basis to landowners hosting wind turbines. In addition, wind projects often contribute significantly to the local tax base and may be sources of local jobs and business activity. Moreover, projects with shares of local ownership can result in additional cash flow back to individuals in the host communities.

The combination of positive and negative impacts on humans and their quality of life presents significant challenges for those attempting to plan for and implement wind energy development. Moreover, many of the widely recognized benefits of wind projects (e.g., renewable fuel, low or zero emissions) accrue at a regional or even global level, while many of the less desirable externalities (e.g., landscape changes, industrial development, sound) are felt in the immediate vicinity of a project.

Source: Kern County, California

Local Economic Impacts

Wind energy projects often provide significant sources of new income for local landowners and governments, and they may generate increased activity for local businesses. Property tax payments on the order of $7,000/MW and landowner lease payments on the order of $3,000/MW to $4,000/MW per year are not uncommon. Routine operations and maintenance work can often be carried out by locals. However, because these types of economic development impacts are not always distributed widely within a given community, concerns of social and economic justice may surface. Nonparticipating project neighbors may experience many of the negative externalities associated with wind projects but receive little or no direct economic benefit.

Mitigating Concerns Around Local Economic Impacts. The uneven distribution of economic development throughout a community can be mitigated in a variety of ways. Community wind projects—projects that incorporate local residents as actual investors or shareholders—have been observed to increase public support (e.g., Jones and Eiser 2009; Zoellner et al. 2008; McLaren 2007; Devine-Wright 2005). In addition, good-neighbor payments and independent community funds have emerged as mechanisms that can help more equitably distribute a project's economic development benefits. Good-neighbor payments provide compensation to individuals who live in

APPALACHIAN STATE UNIVERSITY AND WATAUGA COUNTY, NORTH CAROLINA

Watauga County, North Carolina, population 51,709, lies in the heart of the Blue Ridge Mountains in the northwestern corner of the state. Its history of wind energy stretches back to 1978, when NASA, the U.S. Department of Energy (DOE), and General Electric partnered to install a 240-foot, 2 MW turbine on Howard's Knob, a peak overlooking Appalachian State University. Known as the MOD 1 Program, this turbine was used to generate electricity and gather data until 1982, when funding for the project expired. DOE offered the turbine to a local utility, which declined; with national interest in alternative energy waning, MOD 1 was disassembled in 1983.

That same year, the North Carolina Legislature passed the Mountain Ridge Protection Act of 1983 in response to intense public outcry regarding a condominium project developed on a mountain peak. Known informally as the North Carolina Ridge

of the last four years, and 70 percent of western North Carolina residents surveyed in 2010 felt placing wind turbines atop mountain ridges should be encouraged or at least allowed.

Though there have not yet been any large wind farm applications to test Watauga County's interpretation of the law, activity in this area has been sparked by Appalachian State University. Scanlin had started measuring the wind in 1984 upon his arrival at the university and confirmed good wind resources on mountain ridgetops. In 2004, Scanlin received funding from the North Carolina State Energy Office to establish a small wind-turbine research facility in the county. Realizing the importance of a policy test case, Scanlin mobilized a wind working group to begin wind energy outreach and technical assessment and approached Watauga County about developing a wind energy ordinance to give the project legitimacy and help regulate future development

Source: Appalachian State University

Law, the act limits the height of development on top of protected ridges—those at least 3,000 feet in elevation and at least 500 feet above the floor of the adjacent valley—to no more than 35 feet above the ridge crest.

According to Dennis Scanlin, professor at Appalachian State University and staff member of the university's Energy Center, the Mountain Ridge Protection Act has been one of the biggest barriers to wind energy development in North Carolina. The act included exemptions for electrical transmission towers and windmills, but left the definition of "windmill"—and clear interpretation of the act's applicability to wind energy development—open to debate. The state attorney general has provided an informal opinion that though a solitary windmill would not violate the act, a wind farm would. The current interpretation of the act leaves such decisions to each county. Scanlin notes that elected officials have not yet been willing to clarify this issue to facilitate wind energy development, though statewide interest is growing: wind permitting bills have been introduced in the state legislature each

of wind energy systems. The university provided technical support while the county developed the regulatory language. The entire process took several months. Participants included the planning board, the Blue Ridge Parkway Division of Resource Planning and Professional Services, the National Park Service, citizens with general interest, and property owners who sought to take advantage of residential-scale wind energy. The ordinance—the first of its kind in North Carolina—was passed in 2006.

The Watauga County ordinance addresses both small and large wind energy systems. Small wind energy systems, defined as a single turbine with a rated capacity of not more than 20 kW or multiple turbines on agricultural farms for on-site consumption, are permitted by right subject to requirements. Turbine height is limited to 135 feet, and the ordinance specifies setbacks of one times turbine height from property lines and 1.5 times turbine height from inhabited structures; a building permit is also required. Large wind energy systems, defined as systems of one or more turbines with a rated capacity of more than 20 kW, require a site

(continued on page 45)

the vicinity of the turbines but who do not have a turbine on their property and therefore do not receive landowner lease payments. Community funds are typically initiated by the project developer and administered by an independent authority. Such funds may be designed to finance all types of local community activities but might entail support for energy efficiency improvements, educational fairs or events focused on energy, or reduced electricity rates for low-income residents, among other opportunities. Working with local economic development officials and others to competitively position local service providers, contractors, and businesses will help increase the local distribution of wind energy's economic development benefits.

The mitigation strategies suggested here likely require efforts beyond those of planners to be fully integrated into project and community development. However, by working with policy makers and community members to understand and specify the conditions under which wind energy projects may be authorized, planners can play a critical role in shaping broader local and regional policy and encouraging the types of policies that will facilitate the future planning and development of wind energy.

Aesthetics

For many individuals, landscape, aesthetics, culture, and sense of place are critical variables influencing acceptance and support of wind energy projects (Pasqualetti 2002). People often have deep links to specific landscapes and may be bothered by changes to them (Short 2002; Pasqualetti 2002; Wustenhagen et al. 2007; Wolsink 2007; Firestone et al. 2009). Compared to other renewable energy technologies, wind energy technology is relatively visible; moreover, it is often installed on high points in the landscape (to capture the best wind resource) and in rural regions sometimes noted particularly for their scenic or aesthetic values. Modern wind turbines are also dramatic new presences in the landscape and for some are symbols of industrial development. As the industry continues to mature, wind turbine rotors are expected to grow larger to capture economies of scale, and wind turbine towers are expected to grow taller to capture more productive winds. Both of these features are likely to make wind turbines more visible as well as visible over greater distances. Such trends are likely to increase the aesthetic impact of individual wind turbines, but by reducing the number of wind turbines necessary to produce a given amount of energy and increasing spacing between individual turbines, it may be possible to reduce the overall aesthetic impact of a given project or the industry in general.

Mitigating Aesthetic Impacts. When aesthetic concerns, which are common to many types of infrastructure projects, are not addressed early on at the community level, conditions become ripe for misinterpretation of information, diminished public confidence, and increased project costs, if extensive delays or significant project reconfigurations become necessary. The role of timely and thorough planning is fundamental to minimizing local concerns over aesthetics as well as actual aesthetic impacts. Planning is a primary pathway by which specific areas may be designated for development or set aside for conservation. Areas with particularly high cultural or aesthetic value need to be identified and protected, both from wind energy and other industrial infrastructure, while areas more amenable to landscape alteration should be identified for development.

In instances where federal funds are used or where projects are on federal lands, the process for determining potential aesthetic impacts is relatively well defined under the National Environmental Policy Act (NEPA). In other jurisdictions, clear definitions of geographic scope and video or photo simulations have been required. Early engagement of nearby property owners and residents, effective and accurate forecasting of a project's visual impact,

(continued from page 44)

permit. The ordinance spells out detailed permit requirements, including a site plan, an extensive analysis of potential impacts and proposed mitigative measures, and construction and maintenance plans, and it includes the National Park Service in reviewing and commenting on proposed sites within the Blue Ridge Parkway viewshed. A public hearing is required, and the ordinance provides considerations for the planning board's decision-making process.

Since 2006, there have been four wind energy applications, all for small systems. Appalachian State has also joined in; students voted to increase their fees in 2004 and again in 2007 by five dollars per semester to help fund renewable energy projects on campus. Funds from this initiative, along with contributions from the senior class of 2009 and the local utility company, enabled the university to install a 100 kW wind turbine, the largest in the southeastern United States. Completed in 2009, the turbine provides renewable power for the campus.

For more information:

- Appalachian State University–North Carolina Wind Energy. Available at http://wind.appstate.edu.

- Appalachian State University Renewable Energy Initiative. Available at http://rei.appstate.edu.

- Appalachian State University News. 2009. "Appalachian Installs Wind Turbine on Campus." *University News,* June 24. Available at www.news.appstate .edu/2009/06/24/wind-turbine-on-campus.

- North Carolina Mountain Ridge Protection Act of 1983 (Article 14, North Carolina State Statutes). Available at www.cals.ncsu.edu/wq/lpn/statutes/nc/mountainridgeprotection .htm.

- Watauga County Ordinance to Regulate Wind Energy Systems (2006). Available at www.at.appstate.edu/documents/Watauga Countywind ordinance.pdf.

and factual discussions with landowners and community residents are essential to maintain public trust around a potential project. Utilizing similar turbine types to create consistency and uniformity within a project, selecting turbines of higher generating capacity to require fewer turbine installations for a given energy output, and placing as much electrical infrastructure below grade as possible may also help to minimize aesthetic concerns (Hohmeyer et al. 2005).

Sound

Along with visual aesthetics, nuisance-related impacts are also associated with wind energy projects. The predominant nuisance concern, frequently noted in the media and among opposition groups, is the sound produced by wind turbines. Advances in technology have significantly reduced mechanical noise generated primarily by the gearbox and generator, but aerodynamic noise resulting from the passage of the wind over turbine blades remains significant. Environmental and community noise policies at state and local levels have been developed to protect the public from acute health impacts (McCunney and Meyer 2007). However, existing community noise policies have proven to be insufficient to avoid all noise complaints from individuals who live in the immediate vicinity of wind turbines. In fact, noise complaints have emerged as a persistent problem around the world (Huber and Horbaty 2010). In addition, some neighbors have claimed to have experienced acute health impacts from wind turbine noise including internal pulsing, jitteriness, nervousness, anxiety, nausea, chest tightness, and tachycardia (Pierpont 2010). Aside from a limited number of case studies, however, there is no epidemiological evidence of such health effects (Colby et al. 2009; CMOH 2010; NMHRC 2010). Moreover, it has been noted that many of the symptoms observed in the few case studies that exist are, in actuality, common stress symptoms, which could potentially be induced by annoyance or other factors (Colby et al. 2009).

Unlike other intermittent environmental noise sources such as rail and air traffic, wind turbines appear to be somewhat more annoying at even very modest levels (Pedersen and Waye 2007; Pedersen et al. 2009). In specific conditions where background sound levels are low and wind shear is high, even higher levels of annoyance may persist (Van den Berg 2004, 2008). It has been suggested that increased sensitivity to wind turbine noise may be the result of specific components of that noise that are uncommon in other sources of community noise, such as audible pulsing, which has been shown to increase annoyance from sound generated by locomotives (Kanteralis and Walker 1988). At the same time, research has shown correlations not just between noise annoyance and sound level but also between noise annoyance and unrelated factors including prior attitude toward wind turbines, the visibility of the turbines, and whether or not individuals receive direct financial payments from a project (Pedersen and Waye 2007; Pedersen et al. 2009).

Mitigating Sound Impacts. Perceptions of noise and annoyance are highly subjective. However, the industry has sought to reduce noise emissions from turbines—specifically around the mechanical noise derived from the generator and gearbox. Research has also continued into alternative blade designs to reduce sound emissions (e.g., Lutz et al. 2006; Berg and Barone 2008; Barone and Berg 2010). Nevertheless, it is unlikely that technological solutions will ever completely eliminate noise from wind turbines. In this context, proper planning and siting are critical to minimizing noise impacts on host communities.

In this regard, predictive noise propagation models continue to advance and improve, allowing regulators and potential project neighbors to better understand the levels of noise they are likely to experience. Use and continued refinement of these models is likely to be increasingly important as projects edge closer to densely populated areas. Additionally, sophisticated sound regulations can help to address those specific conditions under which annoyance is likely to be greatest (i.e., periods of high wind shear when there is little ground-level wind to mask wind turbine noise). Such regulations can also place upper bounds on the level of noise or the change from ambient noise resulting from wind energy facilities (Bastasch et al. 2006). Establishing generic setbacks between turbines and property lines or buildings may also allow for sufficient noise mitigation. Project developers may also offer soundproofing for residences that are particularly close to wind turbines.

Shadow Flicker

Another potential nuisance concern is shadow flicker, which occurs from the motion of shadows cast by the rotating blades of a wind turbine. Under some circumstances, shadow flicker is not unlike a low-frequency strobe light. According the Epilepsy Foundation, shadow flicker is not at a high enough frequency to trigger photo-sensitive epilepsy; however, it may be bothersome for individuals subjected to it for extended periods of time. As turbines increase in height, their shadows also grow, increasing the need to address shadow flicker concerns.

Mitigating Shadow Flicker. Shadow flicker can generally be predicted with the use of computer models, and its occurrence on places of human habitation can often be avoided by careful siting of wind turbines combined with strategic placement of vegetation. In extreme cases, sometimes occurring at very northern or southern latitudes during the winter, turbines can be curtailed at specific times of day to reduce annoyance levels for neighbors (Hohmeyer et al. 2005). Although shadow flicker is one of the more easily resolved nuisance challenges, early communication of this concern to project neighbors who might be affected and how it will be addressed is important. Most wind farm modeling software tools have features that facilitate communication and mitigation of shadow flicker.

Obstruction Lighting

All structures over 200 feet in height and structures shorter than 200 feet in height near airports require Federal Aviation Administration (FAA) notification and the installation of obstruction marking lights. Historically these lights can be clearly visible at ground level even at great distances. Thus, obstruction lighting has also been a concern and a potential source of annoyance for nearby residents.

Mitigating Impacts of Obstruction Lighting. Improvements in lighting and lens technology in the last five years have significantly reduced ground-level annoyance. Similarly, the industry has worked to synchronize lights across wind plants, to coordinate with the FAA to minimize the number of lights required within a wind plant, and to include aircraft proximity radars, which allow lights to be turned off in the absence of air traffic. Communities should be aware that options for the lighting of wind projects exist and know what options are available for the specific wind project that affects them. Planners may work with the industry to assist policy makers and regulators in developing appropriate rules, regulations, and technology specifications for wind energy projects within their jurisdiction.

Property Values

Another concern about wind energy facilities is the potential reduction in property values near the turbines. The idea that property values could fall is not inconsistent with what has been seen in other industrial contexts. Development of conventional power plants and transmission lines have resulted in reductions in nearby residential property values (Simons 2006). While there have been few detailed studies of this in specific relation to wind energy facilities, published work has found no evidence of widespread reductions in property value (Sims and Dent 2007; Sims et al. 2008; Hoen et al. 2009). This may suggest that industry siting and setback practices are adequately protecting property owners. (As a comparison, properties near transmission lines see drops in value within a short distance of the lines, but the effect fades after about 100 meters [Des Rosiers 2002].) Alternatively, traditional statistical tools may be unable to identify property value loss among homes that are proximate to wind energy facilities because it is either too infrequent or of too little magnitude.

Source: Kern County, California

Additional research has been suggested, focusing primarily on homes within one mile of wind turbine sites. Such work is expected to provide greater insights into the impact of turbines on property values as the distance between turbines and homes becomes very small and into the distribution of impacts over time. The latter is important to determine whether there are specific periods of significant concern (i.e., during development and construction) and whether concern diminishes as local residents become more accustomed to living adjacent to wind turbines.

Mitigating Property Values Concerns. While there is a variety of potential mechanisms to mitigate property value concerns, there is little consensus within the industry on what is ultimately necessary. In addition, because property value impacts are somewhat derivative of other nuisance issues, mitigation of property value concerns generally entails use of all nuisance-specific mitigation strategies. Nevertheless, frequently noted mitigation strategies for property value concerns include mandating generous setbacks from homes, issuing good-neighbor payments, providing property value guarantees or protection plans, using screening objects including strategically placed vegetation and trees, and adding soundproofing to homes.

Other Factors Influencing Project Success and Concerns

Along with wildlife, aesthetics, and nuisance concerns, some communities and individuals have raised issues related to procedural design and to the perceived justice (or injustice) of the development and mitigation process. Open, transparent communication and participatory development are highly valued by local stakeholders and are most likely to bring about project success (Jones and Eiser 2009; McLaren 2007; Zoellner et al. 2008; Wolsink 2007). Moreover, projects with high levels of public participation are more likely to achieve success and generate a stable network of local project supporters (McLaren 2007). These same themes of open, transparent, and participatory processes are also of utmost importance during planning and when identifying areas to be excluded from or opened up to development (Wolsink 2007; McLaren 2007).

Distance from the project and time are also likely to influence the level of concern individuals have over a given project. With regard to distance, the evidence is somewhat mixed, with some studies (e.g., Warren et al. 2005; Simon 1996) observing increasing acceptance with proximity and among those who are more familiar with wind turbines. In contrast, Van der Horst (2007) and Swofford and Slattery (2010) found that opinions about wind energy may be lower among those living in the immediate proximity of a wind facility. Over time, the evidence appears to suggest that perceptions of projects become more positive and concerns diminish as project neighbors become more accustomed to living in and among a wind energy project (Warren et al. 2005).

Impacts to radar systems are another source of public concern and present a planning challenge for wind energy. Wind turbines can sometimes affect reception and detection of radar signals, making them difficult to distinguish from aircraft or weather and ultimately resulting in security and safety concerns for civilian and military aviation (Krug and Lewke 2009). The primary stakeholders affected by radar issues include the FAA, the Department of Defense, and the Department of Homeland Security (AWEA 2008a). Issues with radar are often the result of dated technology. In some cases, updated software can resolve issues, while others may require new hardware as well (Brenner et al. 2008). Technological solutions such as "stealth" blades may be another alternative (Matthews et al. 2007). Regulatory solutions focused on aviation equipment may also offer some potential to mitigate radar-related concerns among users and operators of radar. Despite the potential for various solutions, coordinating an accepted and adequate response among the stakeholders is expected to take some time. In the near term, planners seeking to identify specific areas for wind deployment are likely to be best served by avoiding those areas where radar conflicts exist.

Conclusions

Siting and planning for wind energy development is a significant challenge, and concerns are wide-ranging and often subjective. In addition, there are a number of variables that planners may not be able to influence or control for. Nevertheless, proactive planning can help communities and regions begin to prepare for wind energy and tackle these issues in a constructive manner. Moreover, planners must be capable of communicating the issues around wind energy and developing solutions that are appropriate for each context and community.

Planning for wind energy projects ideally begins early and entails transparent fact-based discussion, with as much engagement from local residents as possible. In addition, it is important for local planning authorities to understand and communicate why wind energy is necessary and how it fits into the broader energy portfolio in the state or region. Discussion of impacts is most successful and productive when it includes discussion of positive and negative attributes alike. Clear and honest communication and education can go a long way toward mitigating local wind power concerns. Such a multi-stakeholder, transparent, and participatory process is fundamental because the integrity of process is as important as specific substantive issues. Nevertheless, developing workable solutions to individual concerns is also important. For this reason, planning efforts should not be a one-time endeavor, policies should remain flexible so that wind turbine standards can incorporate new information, and planners should work to stay abreast of the most recent experiences emerging from the wind industry.

Source: Kern County, California

WIND CONCERNS SUMMARY

Issue: Wildlife

Concern: Wind turbines kill birds and bats and fragment habitat.

Research: Wind turbines generally pose risks to individuals, not populations. With respect specifically to birds, studies have shown that wind turbines are responsible for far fewer fatalities than other human infrastructure and activities (e.g., vehicles, windows, communications towers, pollution, and house cats). In the case of bats, more data are needed, but concern has emerged primarily from a limited number of exceptionally high fatality events. There is little wind-specific scientific research to either substantiate or refute concerns relating to habitat fragmentation. Improved project planning and siting has significantly reduced wind energy's wildlife impacts, however, and to ensure compliance with state and federal wildlife policies, the wind industry continues to fund scientific research with the purpose of further minimizing wind project impacts to wildlife and habitat.

Issue: Land requirements

Concern: Wind energy will overtake the landscape and prevent otherwise productive use of land.

Research: A modern 2–3 MW turbine typically requires between 70 and 130 acres for siting, but in practice this has ranged from 22 to 250 acres per turbine. Total land use for individual projects is ultimately determined by turbine spacing requirements and local siting constraints. The vast majority of this land can continue to be used for agriculture, ranching, or other uses; only about 3 to 5 percent of the overall footprint of a wind energy facility is occupied by turbines, roads, or other infrastructure and removed from service.

Issue: Emissions savings

Concern: Emissions reductions from wind energy are not real or are significantly less than advocates claim.

Research: Even when considering life-cycle emissions (i.e., emissions resulting from extracting and producing raw materials, manufacturing and transporting equipment, and installing, operating, and decommissioning the facility), emissions impacts from wind energy are significantly lower, in some cases as much as an order of magnitude lower, than conventional fossil fuel life-cycle emissions. The integration of variable output wind energy into the grid system may result in efficiency losses and increased fuel consumption for some existing generation; however, for power systems that have as much as 20 percent of their electricity supplied by wind these efficiency losses have been estimated to offset only a small fraction (roughly 0–7 percent) of the overall emissions savings of wind energy.

Issue: Socioeconomic impacts

Concern: Electricity rates will go up, and economic benefits will not be realized by the local community.

Research: Over the past five years, contracted power prices for wind energy have been within the ranges observed in wholesale power markets across the country. Integrating wind energy into the power system has shown to be modest in cost (on the order of 10 percent) even for power systems that have 20 to 30 percent of their electricity coming from wind. Rate impacts in states with renewable portfolio standards, which often require solar power as well, have generally been limited to 1.5 percent or below.

Property taxes, landowner lease payments, and local operations and maintenance jobs are the primary forms of direct community-based economic benefits. Tax and lease payments are frequently on the order of $3,000 to $7,000 per MW. Good neighbor payments and community funds are other vehicles that can help economic benefits from wind energy projects flow throughout the community. Wind farm development can also lead to jobs for local service providers, contractors, and businesses, resulting in increased economic development for local communities. When local residents directly invest in projects (e.g., community wind) local economic development is further enhanced by the return of project profits to the host community.

Issue: Aesthetics

Concern: Wind turbines will ruin views and harm the integrity of the cultural landscape.

Research: Wind energy facilities are often quite visible; however, responses to wind power are highly subjective. Some see wind energy as a symbol of technological advancement and energy independence, while others see only new industrial development. Techniques exist to accurately portray and evaluate aesthetic impacts; such visual impact analyses can help local residents better understand exactly what a project might look like. Proper planning and avoidance of specific highly valued aesthetic areas is also important.

Issue: Sound

Concern: Sound from wind turbines is annoying and will disrupt sleep; it may also cause health impacts.

Research: Wind energy facilities, like other land uses, are subject to state or local environmental noise guidelines designed to protect individuals from acute noise-related health impacts. In addition, there is no widespread evidence of acute health impacts from wind energy sound in the literature. At the same time, individuals living in close proximity to wind turbines may find the noise they generate annoying, and in some cases it can disrupt sleep. While those individuals that report annoyance are in the minority, proper siting and planning are critical in mitigating potential project noise issues.

(continued on page 52)

(continued from page 51)

Issue: Shadow flicker

Concern: Shadow flicker is disruptive and annoying and prevents reasonable enjoyment of the land in and around wind projects.

Research: Persons subjected to shadow flicker for long periods of time are likely to find it bothersome. However, shadow flicker is easily modeled and can generally be avoided with proper planning and siting. Because of the ease with which shadow flicker can be predicted and ultimately mitigated, many localities specifically address shadow flicker in local ordinances.

Issue: Obstruction lighting

Concern: Obstruction lighting will be highly visible and generate a significant nuisance for the community.

Research: Improvements in lighting and lens technology have significantly reduced the ground-level impacts of wind turbine obstruction lighting. The industry continues to develop new technology to minimize disturbances from obstruction lighting, including proximity sensors that allow the lights to be turned off when no aircraft are in the vicinity.

Issue: Property values

Concern: Wind projects will negatively affect property values.

Research: There has been little persuasive evidence to date of widespread property-value impacts from wind turbines. However, because such trends are not unprecedented in other industrial contexts and effects sometimes fade very quickly (e.g., transmission lines), additional research focusing specifically on homes in very close proximity to wind turbines is needed.

Issue: Radar and electromagnetic interference

Concern: Wind turbines will interfere with television signals and could potentially jeopardize safe aviation navigation.

Research: In some cases, wind turbines can cause electromagnetic interference, but these problems can largely be resolved with preconstruction analysis and proper siting. Impacts on military and aviation radar present more significant problems. Generally, however, turbines are not allowed to be built in locations where they may potentially affect aviation safety. Though a variety of technical and regulatory solutions to radar interference exist, coordinating an acceptable solution with the multiple federal agencies involved is likely to require some time. For the time being, therefore, avoid wind energy development in areas where conflicts with aviation or military radar systems are expected. ◀

CHAPTER 4

Regulatory Environment

 State lawmakers have allocated authority to regulate the siting of wind facilities in different ways. In more than half the states, local governments are solely responsible for regulations that govern the siting of all wind power–generating facilities, most often using their land-use planning, zoning, and related permitting authorities. Nearly all the remaining states also give local governments substantive roles in small and noncommercial wind facility siting decisions, or make local government approval necessary but not solely sufficient to meet state requirements. A significant minority of states confer upon state siting bodies or commissions concurrent, preemptive, or exclusive authority over siting larger generating facilities.

KANSAS

Kansas, which has more than 1,000 MW of installed wind-generating capacity, is one state where local governments are exclusively responsible for siting all types of wind energy–generating facilities. This means that local governments can decide to reject wind power as a permissible land use. For example, Waubansee County, an 800-square-mile county in the scenic Flint Hills, adopted an initial moratorium on conditional use permits for commercial wind energy facilities in order to consider revisions to its land-use plan and zoning ordinance. In 2004, the county board of commissioners adopted an updated comprehensive plan and revised zoning ordinance. While the county's planning commission had recommended allowing commercial wind facilities as a conditional use, the board instead adopted an ordinance prohibiting commercial wind projects, defined as facilities able to generate 100 kW or taller than 120 feet, or projects with multiple turbines. Landowners and wind-rights holders filed suit, and in 2009 the Kansas Supreme Court upheld the county zoning ordinance, finding that the board's decision to prohibit commercial wind was within its legislative discretion and that it was reasonably supported by the record. The court noted that a total ban might be "unwise" but was not illegal and that the county's action was not expressly nor impliedly preempted by state law (*Zimmerman v. Bd. of County Comm'rs*, 289 Kan. 926, 218 P. 3d 400 [2009]).

STATE AND LOCAL GOVERNMENT AUTHORITY OVER WIND FACILITY SITING

Jim McElfish and Sara Gersen

States have chosen different mechanisms for controlling local regulation, different thresholds and ceilings for the size of facilities subject to such regulation, and different priorities for the content of siting regulation (McElfish and Gersen 2011). This has led to a tremendous variety in local government roles in wind energy siting decisions. (See Table 4.1, opposite page.)

Local Siting

In the most common approach to wind facility siting, local governments are solely responsible for siting wind facilities, and state law does not limit their regulatory power. (See sidebar.) State land-use planning and zoning enabling laws allow local governments to engage in land-use regulation, which includes the regulation of wind facility siting. This governance model is the default situation in states that do not have utility siting boards or legislation specific to wind facility siting. Even where state governments take an active role in permitting wind facilities, local governments usually have exclusive control over the siting of small (and sometimes medium-sized) wind facilities, which typically do not fall under state jurisdiction.

This model of regulation gives local planners and governing bodies the most freedom in promoting local interests. It allows local governments to limit wind energy facilities to particular zones and locations and to prescribe conditions for their construction, operation, and decommissioning—or to prohibit them altogether. Local decision makers who strongly support wind power development can facilitate rapid project approval by establishing clear rules, standards, and procedures for such facilities. By the same token, local officials in states that allow local control may find it very difficult to deal effectively with wind energy proposals if such local regulations and guidance have not been established. Communities should be proactive in introducing wind energy into the comprehensive planning process and developing local regulations that clearly define where and in what manner wind energy systems are appropriate. Local governments operating under this model of autonomy will need to develop expertise (either on their professional staffs or by engaging consultants) to draft ordinances, evaluate applications, and devise and apply appropriate permit conditions.

Additional regulations may also apply in states where local governments have jurisdiction over siting. For example, in some states, a state public utility commission or board must make determinations about new generating capacity or ability to connect to the power grid—decisions that will indirectly affect wind facility siting. The state may authorize electricity production through a "certificate of need" without considering land use or siting issues. For example, a commercial energy-generating facility cannot be built in North Carolina unless the Utilities Commission issues "a certificate that public convenience and necessity requires, or will require, such construction" (N.C. Gen. Stat. § 62-110.1). The commission issues certificates based on the need for new supplies of electricity and the anticipated costs of the proposed facilities, while siting permits remain the responsibility of local governments. In addition, state boards and commissions are often responsible for transmission siting policies, which may drive developers to build facilities in particular locations. In areas with significant wind resources and current or planned transmission lines, planners should prepare for significant interest from wind energy developers.

Where particular resources—such as wetlands, state waters, wildlife protection areas, or threatened and endangered species—are affected, state environmental permitting or other reviews may apply in addition to local

TABLE 4.1: STATE AND LOCAL GOVERNMENT AUTHORITY OVER WIND ENERGY FACILITY SITING

State	Local siting authority	Local siting authority plus special constraints	Dual permitting	State siting
Alabama	All			
Alaska	All			
Arizona	All			
Arkansas	All			
California	Over 50 kW	Under 50 kW		
Colorado	Net-metered		All but net-metered[a]	
Connecticut	Under 1 MW		Over 1 MW[a]	
Delaware	All			
Florida	All			
Georgia	All			
Hawaii	All (but dual in some use districts)		State Land Use Commission—some districts	
Idaho	All			
Illinois	All commercial-scale	End-user systems		
Indiana	All			
Iowa	Under 25 MW		Over 25 MW	
Kansas	All			
Kentucky	Under 10 MW		Over 10 MW	
Louisiana	All			
Maine	Projects under 20 acres in municipality		Projects over 20 acres in municipality	Land Use Regulation Commission sites in unorganized areas
Maryland	Under 70 MW		Over 70 MW, in general	
Massachusetts	Under 100 MW			Over 100 MW[b]
Michigan	All			
Minnesota	Under 5 MW	Counties may permit 5–25 MW if state delegates		Over 25 MW; 5–25 MW if not delegated to county[c]
Mississippi	All			
Missouri	All			
Montana	All			
Nebraska	Under 2.5 MW			Over 2.5 MW
Nevada	All commercial-scale	End-user systems		
New Hampshire	100 kW–30 MW	Under 100 kW		Over 30 MW[c]
New Jersey	All	End-user systems; and wind power eligible for variance as "inherently beneficial use"	State permit or permit by rule needed in coastal zone	
New Mexico	Under 300 MW		Over 300 MW[a]	
New York		Under 25 MW, restrictions must be required by law		Over 25 MW [a, c]

(continued)

TABLE 4.1: *(continued)*

State	Local siting authority	Local siting authority plus special constraints	Dual permitting	State siting
North Carolina	All			
North Dakota	Under 60 MW			Over 60 MW[b]
Ohio	Under 5 MW			Over 5 MW
Oklahoma	All	Decommissioning required		
Oregon	Under 35 MW, unless developer election			Over 35 MW, and where developer of smaller facility elects [a,b]
Pennsylvania	All		All but net-metered[a]	
Rhode Island	Under 40 MW			Over 1 MW[a]
South Carolina	All			
South Dakota	Under 100 MW	Setback required	Over 100 MW	
Tennessee	All			
Texas	All			
Utah	All			
Vermont				All[c]
Virginia	Under 5 MW		Department of Environmental Quality permit-by-rule for 5–100 MW; over 100 MW, State Corporation Commission siting	
Washington	All, unless developer elects state review	Deferential review in energy overlay zones		All, with developer election[a]; expedited siting when consistent with local plans and ordinances
West Virginia	End-user systems		All commercial-scale	
Wisconsin	Under 1 MW	1–100 MW		Over 100 MW
Wyoming	Under 0.5 MW	Over 0.5 MW but fewer than 30 turbines	30 or more turbines	

[a] State may preempt local decisions in limited circumstances.
[b] Incorporates local standards
[c] Must consider local plans and ordinances

land-use decision-making processes. Some states (e.g., California, Washington, Hawaii, Massachusetts, New York, and Montana) also apply environmental impact review requirements to state or local permitting and siting.

A number of state bodies or associations have developed handbooks, model ordinances, or guides to assist local governments in developing wind energy regulations. Such resources have been developed by state bodies in Kansas, Maine, Massachusetts, Michigan, New Jersey, New York, Oregon, Pennsylvania, South Dakota, Utah, and Wisconsin, and by other organizations or institutions in Illinois, Minnesota, and North Carolina. (See resources list.)

Local Siting within State-Defined Constraints

In states where local governments issue the only siting permits for wind energy facilities (or for certain sizes of wind facilities), legislatures have taken a variety of approaches to circumscribe local siting authority. Some have focused on protecting wind energy systems from local regulations that they view as especially burdensome, especially small systems under defined capacity thresholds and systems that are used for on-site power by the landowner (end-user systems) or that allow net metering.

For example, Nevada's enabling laws prohibit local governments from adopting "an ordinance, regulation or plan or tak[ing] any other action that prohibits or unreasonably restricts" end-user systems. Nevada defines an unreasonable restriction as one that significantly decreases the system's performance or efficiency without allowing a comparable alternative. Regulations concerning height, noise, safety, or FAA compliance are permitted regardless of whether they meet this test (Nev. Rev. Stat. Ann. § 278.02077). Similarly, New Hampshire prevents municipalities from adopting regulations that "unreasonably limit" the installation or performance of small systems that generate fewer than 100 kW and produce energy for on-site consumption (N.H. Rev. Stat. Ann.§§ 674:62–66). New Hampshire's definition of unreasonable limits is much broader than Nevada's, however: it prohibits municipalities from applying noise limits under 55 dBA, generic height restrictions that do not specifically address wind systems, and setbacks of more than 150 percent of system height.

California has taken a slightly different approach to promoting small wind energy systems. The state allows counties to regulate systems that generate fewer than 50 kW, but local ordinances may not be more restrictive than a set of conditions specified in the law. For instance, setback requirements may not be greater than the height of the system (Cal. Gov. Code §§ 65893–99).

Illinois gives incorporated municipalities almost total control over wind energy siting within their jurisdictions, which include adjacent unincorporated areas within 1.5 miles of municipal boundaries. County enabling laws explicitly allow counties to establish standards for wind facilities, including height restrictions and restrictions on the number of turbines in a given area. However, setback requirements for end-user systems cannot be greater than 110 percent of system height (55 Ill. Comp. Stat. 5/5-12020; 65 Ill. Comp. Stat. 5/11-13-26).

Several other states have focused on limiting local regulatory action that might impede the development of medium or large wind facilities. Wisconsin has developed but not yet implemented the most comprehensive and detailed limitations on local decisions affecting wind siting. In 2009, newly updated state legislation ordered the state's Public Service Commission (PSC) to develop rules for siting facilities with generating capacities from 1 to 100 MW, addressing issues such as setbacks, noise, flicker, decommissioning, and application procedures, with the assistance of a Wind Siting Council composed of a variety of stakeholders (Wis. Stat. § 196.378(4g)). Previously, local governments had limited authority to regulate wind siting, and state judicial decisions had cast further doubt on their ability to prescribe local standards. Under the 2009 legislation, local wind energy ordinances cannot be more restrictive than the new PSC rules (Wis. Stat. § 66.0401(4)(g)). However, in March 2011 the Wisconsin legislature suspended the PSC rules before they could go into effect, leaving local regulation in uncertain status.

In Wyoming, local permitting for wind facilities with capacities over 0.5 MW must comply with a detailed set of state requirements (Wyo. Stat. § 18-5-501(a)(ii)). To obtain a local permit, applicants must provide emergency management and decommissioning plans, certify that there will be no advertising on the equipment, meet a suite of setback requirements, and comply with other rules (Wyo. Stat. § 18-5-503). A state permit is required for facilities with more than 30 turbines. The state Industrial Siting Council provides technical assistance to help counties evaluate the potentially significant environmental, social, or economic impacts of wind development.

Minnesota offers a one-stop permitting system for wind energy facilities with generating capacities over 5 MW, with the Public Utilities Commission typically responsible for issuing permits and ensuring that applicants comply with both local and state requirements. Minnesota's legislation allows the PUC to delegate authority to the counties to permit facilities up to 25 MW. Counties must apply the PUC's siting standards, but they share the PUC's discretion to grant variances in the public interest. The PUC provides the counties with technical assistance (Minn. Stat. § 216F.08).

Source: Kern County, California

In New Jersey, the legislature, while recognizing that wind projects are subject to local zoning, has made it easier for wind project proponents to get variances from local zoning ordinances. Under New Jersey land-use law, a zoning variance may be granted when zoning boards find (1) "special reasons" exist for the variance; and (2) the variance "can be granted without substantial detriment to the public good and will not substantially impair the intent and the purpose of the zone plan and zoning ordinance" (N.J. Stat. § 40:55D-70(d)). However, for an "inherently beneficial use," which includes "a wind, solar or photovoltaic energy facility or structure," the first requirement is presumed to be met, and the second requirement is met if the project would not cause a substantial detriment to the public good (N.J. Stat. § 40:55D-4).

In Washington, where a developer may choose whether to obtain a siting permit from the local government or the state Energy Facility Site Evaluation

Council, state law encourages counties to create "energy overlay zones," which is "a formal plan enacted by the county legislative authority that establishes suitable areas for siting renewable resource projects based on currently available resources and existing infrastructure with sensitivity to adverse environmental impact" (Wash. Rev. Code § 36.70C.020). If the decision to site a wind facility in these special zones is challenged, a court must uphold the siting decision as long as either the ordinance for the zone is consistent with the state's Department of Fish and Wildlife's guidelines or the county has prepared an environmental impact statement on the overlay zone with mitigation as required (Wash. Rev. Code § 36.70C.130).

Some states have enacted requirements that apply directly to wind facilities, regardless of the permitting or siting authority. In South Dakota, state law imposes a setback requirement on wind turbines taller than 75 feet. These turbines "shall be set back at least five hundred feet or 1.1 times the height of the tower, whichever distance is greater, from any surrounding property line" (S.D. Codified Laws § 43-13-24). Oklahoma state law requires defined decommissioning actions for commercial wind facilities within 12 months of project abandonment or cessation of use (Okla. Stat. tit. 17, §§ 160.12–19).

Dual Authority

In several states, applicants must obtain siting permits from both local and state permitting authorities for larger projects or projects of certain types, while local governments retain exclusive control over smaller facilities. The threshold for state certification varies widely. In Colorado, all commercial power-generating facilities require a PUC permit. In New Mexico, only facilities with generating capacities over 300 MW must seek state approval.

Local and state permitting processes may operate independently, but state permitting may still affect local planners' approaches. The state siting process may create delays in the permitting and approval process that are not present in states where local governments have exclusive jurisdiction. Conversely, state requirements may lighten the load of local governments in their evaluation of complex projects. For instance, under Virginia's state permit-by-rule provisions, applicants must perform prescribed wildlife surveys and mitigate wildlife impacts for wind facilities over 5 MW (Va. Code §§ 10.1-1197.5 et seq.), lessening the need for local regulations to address this issue.

In Hawaii, the state's land-use law empowers the State Land Use Commission (LUC) to categorize all land into one of four districts: urban, rural, agricultural, or conservation. Certain uses in certain districts require state special use permitting as well as local approval; some uses are approved by statute, such as wind energy facilities in agricultural districts (Hawaii Rev. Stat. § 205-2(d)(4), (d)(7)).

In general, the dual-authority approach simply assures that projects will not go forward unless they satisfy both state and local concerns. In a few states with dual-permitting schemes, however, the state siting body can sometimes preempt local decisions or requirements. For instance, if a local government in Colorado denies an application for a commercial wind facility or imposes unreasonable conditions, the applicant may appeal to the state Public Utilities Commission. The commission may override the local decision if it finds that "the conditions imposed by a local government action unreasonably impair the ability of a public utility or power authority to provide safe, reliable, and economical service" (Colo. Rev. Stat. § 40-4-102). In addition, applications for state siting permits must include all relevant local permits, which are automatically deemed approved if they are not issued within 180 days of a preliminary application or 90 days of a final application

LAMAR, COLORADO

The City of Lamar, population 7,804, is located in Prowers County in the southeastern corner of Colorado. It provides a good example of how a smaller wind project can "piggyback" on the development of a larger one nearby. In 2003, Prowers County began construction on the Colorado Green Wind Power Project—at that time the fifth-largest wind farm in the world with 108 1.5-MW turbines—20 miles south of Lamar. Local officials became interested in the project, and the city commissioned its own wind farm in February 2004. The local power utility was able to benefit from economies of scale and local knowledge gathered during the Colorado Green project in order to install and operate several utility-scale turbines for its own use.

The Lamar Wind Energy Project consists of four 1.5-MW wind turbines. Three are owned by local utility Lamar Light and Power (LL&P), and the fourth is owned by the Arkansas River Power Authority (ARPA), a local joint-action agency supplying wholesale energy to its six municipal members. All of the electricity generated by the turbines feeds directly into ARPA's power distribution, where the electricity may or may not be dispersed into Lamar.

Rick Rigel, superintendent of LL&P, believes Lamar's decision to piggyback on the Colorado Green Project was an excellent decision for their small city, as they were able to purchase wind turbines from General Electric (GE) at Colorado Green's group price. Lamar also struck a five-year contract with GE's Wind Energy division for maintenance services. When the contract expired in late 2009, GE was not interested in renewal, so the city was forced to search for another way to maintain their four turbines. Rigel was fortunate in meeting Jim Gill, a past employee of GE who was experienced in wind turbine construction and clean energy. Gill was immediately hired as head of maintenance, and his current team consists of three volunteers from Lamar who are dedicated to learning the skills of turbine upkeep. This benefits LL&P as well as the municipality, as local involvement with the turbines brings a sense of community to the area.

The turbines are located four to five miles from the center of town, on land leased directly from local farmers. The city's zoning regulations have allowed wind energy conversion systems (WECS) since 1986 as uses permitted in any district subject to requirements, including setbacks from lot lines ranging from 100 to 385 feet based on turbine rotor diameter, height limits of 20 feet above maximum district height restrictions, and additional aesthetic and safety standards.

Rigel and Gill report that citizen support has been positive and that landowners lined up to host the turbines. Rigel noted that the high cost of turbines is always an issue, but this did not stop Lamar from moving forward with the process; the city issued 20-year general revenue bonds for $6 million to fund turbine installation. Load balance fees increased initial operating costs, which posed a challenge.

The Lamar Wind Energy Project produces enough electricity to serve about 14 to 15 percent of the city's energy needs, and is working at high capacity with great reliability. Lamar Light and Power is proud of its wind farm, as it stands as proof that small utility companies can in fact own and operate wind energy projects successfully.

For more information:

- Lamar City Code (2000). Chapter 16, Article XVII, Section 16-17-190, Wind Energy Conversion Systems (WECS). Available at http://lamarco.govoffice3 .com/vertical/Sites/%7B56E8FCFD-5F3B-42B1-A514-5555E2C39101%7D/ uploads/%7B2C7BFE48-DD8B-4FBC-A1FE-A5860D05D92D%7D.pdf.

- Lamar Light and Power. "Generation Portfolio." Available at www.lamarlightand power.com/generation_portfolio.html.

- Prowers County Development. "Colorado Green Wind Power Project." Available at www.procolorado.org/html/colorado_green.html.

(Colo. Rev. Stat. § 29-20-108). (For a case study from Colorado, see the sidebar.)

State Siting

In this approach, permitting larger wind energy facilities is the sole responsibility of state authorities, though again, local governments usually retain jurisdiction over smaller facilities under specific thresholds. Within this category, there is great variety in how local planners and decision makers can influence the state siting process. Where state permitting boards are directed by state law to consider or apply local regulations, local planners should take advantage of the opportunity and include clear standards to make sure that community concerns are taken into account by state decision makers. For instance, Rhode Island's Energy Facility Siting Board is responsible for issuing all licenses and permits for energy-generating facilities over 40 MW, but applicants must demonstrate that they comply with all local rules and ordinances. Local officials have the opportunity to issue an advisory opinion on the application of their regulations (R.I. Gen. Laws §§ 42-98-1–11).

Washington's state siting council has authority to approve wind energy facilities that choose to seek approval from the state "regardless of the generating capacity of the project." Approval pursuant to this process preempts other regulation of facility "location, construction, and operation conditions," although the council must include conditions to protect state or local governmental interests, including those that it preempts or supersedes (Rev. Code Wash. § 80.50.100). The Supreme Court of Washington upheld an EFSEC decision that imposed setbacks that were less than those than a county board had required (*Residents Opposed to Kittitas Turbines v. State Energy Facility Site Evaluation Council*, 165 Wa.2d 275, 197 P.3d 1153 [2008]).

TOWN OF FENNER, MADISON COUNTY, NEW YORK

Madison County is home to several successful wind-energy projects. A county of 70,000 residents in upstate New York, it has been involved in clean energy generation since the 1990s. Madison County hosts 43 utility wind turbines on three wind farms and numerous residential turbines. Its largest wind project, spanning more than 2,000 acres of farmland privately leased from fourteen separate landowners, is located in the Town of Fenner, 40 miles southeast of Syracuse. With 19 1.5-MW turbines in use, Fenner Wind Farm produces an estimated 89,000 megawatt-hours per year, enough to power approximately 10,000 households.

A primarily rural town with an agriculture-based economy, Fenner's future-oriented thinking is demonstrated in its comprehensive plan. The plan places the development of alternative agricultural and renewable energy enterprises alongside the preservation of rural character and quality of life as prime concerns of the town's land-use regulatory efforts, and it recommends the consideration of zoning provisions to promote alternative energy initiatives.

According to Town Supervisor Russell Cary, Fenner's road to wind farm development began in 1998 with prodding from a representative from the Atlantic Renewable Energy Company, a large purveyor of wind energy on the East Coast. Studies conducted over two years showed that the area had average wind speeds of 18 mph, enough to sustain wind turbines. A grant from the New York State Energy Research and Development Authority (NYSERDA) helped offset the capital costs of the Fenner project, which was developed by Enel North America and Atlantic Renewable Energy. The project was considered a public utility use and went through the special use permit process to ensure appropriate turbine siting; multiple public hearings and information meetings were held to inform the community about the project and answer questions from residents.

In 2000, Fenner amended its land-use regulations to specifically address wind energy. The amendment designates a list of specific tax map parcels as "District C," in which wind power electricity generation and transmission facilities are permitted as special uses. It further provides special criteria for commercial wind facilities, addressing electromagnetic and telecommunications interference, FAA lighting, noise, and safety issues, and it requires visual impact analyses for projects. The amendment establishes setbacks of 1.5 times structure height from property lines and structures; a provision was added in 2001 to allow reduction of this requirement in cases where two adjacent C District properties both host turbines or where the property owner records a development easement restricting use within a reduced setback area.

The wind farm was completed in December 2001 and operated uneventfully until 2009, when one of the turbines collapsed into an empty field. The wind farm was taken offline for eight months for analysis and repairs. Though engineers were unable to determine the cause of the collapse, no one was hurt, and Enel reinforced the bases on all remaining turbines before resuming operations. No further problems with the turbines have been reported.

Public response to the project has been overwhelmingly positive. Though the turbines are very visible throughout the area and some residents consider them eyesores, Cary notes that wind energy is seen as just another crop in the community, and local farmers are reimbursed for hosting turbines on their properties. In 2001, the Fenner Renewable Energy Education (FREE) Center, a hands-on museum dedicated to green energy, opened its doors. A state-of-the-art facility in the process of becoming LEED certified, it has been a tourist attraction for the community, drawing local schoolchildren and visitors from across the country. Located on a donated plot of farmland two acres in area, it is only 50 yards away from a turbine—the closest one can safely be to a turbine anywhere in the nation without a hardhat.

In December 2010, Fenner received the Outstanding Government Leader Award, part of the Alliance for Clean Energy New York's Sustainable Energy Leadership Awards, for being a leader in green energy. The town takes pride in its wind farm and the priority it places on educating others about renewable energy.

For more information:

- Town of Fenner Comprehensive Plan (2009). Available at www.townoffenner.com/pdf/2010Final_Comprehensive_Plan.pdf.

- Town of Fenner Local Law No. 2000-1 (2000). Available at www.townoffenner.com/permits/LL2000-1.pdf.

- Town of Fenner. Local Law No. 2001-1 (2001). Available at www.townoffenner.com/permits/LL2001-1.pdf.

- "Fenner Receives Outstanding Government Leader Award." 2010. *Madison County New York News*. November 10. Available at www.madisoncountycourier.com/2010/11/10/fenner-receives-outstanding-government-leader-award.

- Fenner Renewable Energy Education (FREE) Center. Available at www.thefreecenter.org.

- "Fenner Turbines Spinning Again After Eight Motionless Months." 2010. *WSYR News*. December 27. Available at www.9wsyr.com/news/local/story/Fenner-turbines-spinning-again-after-eight/Rbs42QIY8EyvGn-inmnl5g.cspx.

- Fenner Wind Power Project Info Sheet. Available at www.madisoncountyagriculture.com/altenergy/FennerProjectinfosheet1.pdf.

New York until 2011 provided substantial leeway to local governments for regulation of wind facility siting, but it limited some of the siting restrictions that local governments might impose. State law expressly preserved the authority of local governments to apply zoning ordinances, building codes, and certain state environmental laws over these facilities, but it precluded local governments from imposing any conditions or requirements not provided by these laws and ordinances (N.Y. Energy Law § 21-106(2)). In 2011, recognizing that this approach had produced varying results for large wind facility siting, the state legislature adopted a one-stop state siting board process for energy facility siting over 25 MW, retaining primacy of local regulation only for smaller facilities. The seven-member New York State Board on Electric Generation Siting and the Environment will consist of five permanent members plus two members appointed for each proceeding from the community where the proposed facility would be located. The law provides that the board will take into account local requirements and consider evidence from local governments supporting such requirements but will preempt requirements it deems "unreasonably burdensome" (Power NY Act of 2011, § 12, codified at N.Y. Pub. Serv. L. § 160 et seq.).

Some state boards are empowered to site large wind energy facilities without regard to local ordinances and plans. One example is in Ohio, where the Power Siting Board has exclusive authority to site wind energy facilities over 5 MW (ORC Ann. §§ 4906.01–4906.20). The legislature did not make local regulation a mandatory consideration in the board's decision making process. Further, the law explicitly states that no other agency or political subdivision may impose any permit requirement or condition on facilities under the board's jurisdiction.

Other state siting boards must consider local plans and ordinances but are not bound to make their decisions consistent with local rules. The Vermont Public Service Board is one of the rare state agencies exclusively responsible for siting all wind energy facilities regardless of size. The board may issue a siting permit only for projects that it determines "will not unduly interfere with the orderly development" of an area (30 V.S.A. § 248). The board must give "due consideration" to, but is not required to comply with, the recommendations of municipal and regional planning commissions and the contents of municipal plans (*In re UPC Vermont Wind LLC*, 2009 Vt. 19 [2009]).

State and local governments have engaged in a great deal of new law making as wind energy at all scales enters the marketplace. State governments have attempted to balance recognition of local concerns against state goals to increase the availability of renewable electric power and reduce greenhouse gas emissions. Many local governments have adopted new ordinances with very little previous exposure to wind technologies, impacts, and needs. Fortunately, planners and elected officials now have a wealth of technical information on which to draw, as well as numerous state and local practices to consult for guidance. But the legal landscape is far from uniform.

TRANSMISSION AND INTERCONNECTION

Kevin Porter and Sari Fink

Transmission

"Transmission" refers to the bulk transfer of electrical power over high-voltage lines from generating plants to substations. This is different from distribution lines that take the electric power from substations to the final consumer. New transmission is particularly important for wind energy. Wind power projects must locate in areas with sufficient wind resources, which often do not have high levels of electricity demand and may not have adequate transmission infrastructure in place. In addition, there is also a mismatch between the relatively short timeframe needed to develop a wind power project compared to the longer timeframe typically required to build new transmission. This section briefly describes the regulation and siting of transmission in the United States. This section also describes how generators, including wind power generators, are interconnected to the electric grid.

Transmission Regulation in the United States. The United States has a complex and multijurisdictional approach toward regulating the electricity industry. Regulation of transmission, and of the electricity industry in general, is divided among the federal government, states, and local governing boards, depending on the context. Federal entities involved with transmission include the following:

- *Federal Energy Regulatory Commission (FERC)*: wholesale electricity markets and interstate transmission and generation rates

- *North American Electric Reliability Corporation (NERC)*: maintenance of electric reliability through issuance of standards, with oversight by FERC

- *U.S. Department of Energy (DOE)*: national interest electric transmission corridors and sales and transmission of electricity to foreign countries, among other things

- *U.S. Department of Agriculture (including the U.S. Forest Service) and the Department of Interior (including the Bureau of Land Management)*: rights-of-way and land-use management on federal land, as well as financing and oversight of federal power marketing administrations

- *Federal Utilities and Power Marketing Administrations*: operation of federal hydroelectric facilities and administration of their transmission network

In general, FERC regulates interstate electric wholesale transactions and determines wholesale transmission and generation rates (i.e., generation and transmission between market participants, not the generation or distribution of electricity to the ultimate consumers). Alaska, Hawaii, and most of Texas are exempt from FERC jurisdiction as they are not linked to the main interconnections in the United States. Another one-third of transmission facilities in the United States are not owned by FERC-jurisdictional entities (typically municipal utilities, power marketing administrations, or rural electric cooperatives) and are not subject to FERC regulation. FERC requires FERC-jurisdictional transmission providers to participate in local and regional transmission planning.

States regulate bundled retail rates (i.e., for generation, transmission and distribution) for entities under their jurisdictions, which typically include investor-owned electric utilities. Some states may also include municipal utilities and rural electric cooperatives. States also have jurisdiction over distribution lines that serve end-use customers directly (Daryanian et al. 2009).

Transmission Siting. States are currently responsible for siting transmission, with the exception of the federal power marketing administrations and the Tennessee Valley Authority, which have their own siting responsibilities. In addition, federal approval from agencies such as the U.S. Forest Service or the Bureau of Reclamation is required if a proposed transmission facility crosses federal land. Also, FERC can step in to approve the siting and construction of transmission projects in the DOE-designated Mid-Atlantic and Southwest Area national interest electric transmission corridors, though it has not yet done so.

CALIFORNIA RENEWABLE ENERGY TRANSMISSION INITIATIVE

The California Renewable Energy Transmission Initiative (RETI), launched in 2007, was a statewide effort by energy stakeholders to plan for the transmission projects necessary to link potential renewable energy projects to the grid. State mandates called for retail electrical suppliers to obtain 20 percent of their energy from renewable sources by 2010, so RETI was formed to help plan for the transmission infrastructure needed to enable rapid development of the state's renewable resource areas. RETI results were intended to point to the highest-priority transmission projects needed to connect renewable energy resources located in remote locations to the statewide high-voltage transmission grid.

RETI was a joint effort among the California Public Utilities Commission (CPUC), the Energy Commission, the California Independent System Operator (California ISO), investor-owned utilities, and public utilities. Its planning objectives were to identify the state's Competitive Renewable Energy Zones (CREZs), determine which ones were priorities for development, refine analyses of those, considering both economic and environmental attributes, and develop a statewide conceptual transmission plan.

Perhaps the most distinctive and important feature of RETI was its focus on collaborative stakeholder involvement and the open, transparent nature of the process. Rich Ferguson, RETI coordinator, emphasizes that the RETI process differed significantly from what had come before: "The usual utility planning process is a very staff-driven operation, creating a document that the outside world then reacts to. RETI was very different; it was stakeholder-driven from day one." The initiative was overseen by a coordinating committee made up of representatives from state regulatory agencies and utility representatives, but the primary working group, the Stakeholder Steering Committee (SSC), comprised representatives of transmission owners and providers; generators; utilities and power purchasers;

(continued on page 65)

Transmission siting is complicated, due to the geographic reach of transmission projects and impacts on the environment, land use, wildlife, and local and regional economies. Each state typically addresses transmission siting in its statutes. After required environmental, socioeconomic, or other studies have been performed, a utility may proceed with construction of a line if no challenges are filed after a specified time period. Generally, the utility must prove that the proposed transmission line meets a demonstrated need (usually for maintaining electric reliability), and the responsible state agency must rule on whether the transmission facility is necessary and whether it meets the public interest (Meyer and Sedano 2002).

The process begins when a utility requests a certificate of convenience and necessity (CPCN; different terms may be used in different states) from the siting agency, generally the state public utility commission, to construct a transmission facility. Although it may vary by state, the CPCN requires information on the proposed transmission project, such as a general description; estimated cost; expected construction date; length of construction; projected in-service date; a map of the proposed location; information on alternatives that were assessed and the reasons why they were eliminated; and justification for the proposed transmission facility. Utilities may also need to demonstrate "prudent avoidance" of certain problems by, for example, minimizing exposure to electromagnetic fields. CPCN applications also may require an environmental assessment that evaluates the potential impact of the proposed transmission line, such as potential disruption of habitat, impact on state- or federal-listed endangered or threatened species, and socioeconomic or visibility impacts. If the proposed transmission project crosses or involves federal lands, then an environmental impact statement or environmental assessment under the National Environmental Policy Act will be required.

Depending on the state, local governments may also have responsibility for approving or denying proposed transmission facilities. They are often given a particular time frame (e.g., 120 days), after which the utility's CPCN application is approved if the local government does not act. Should the local government deny the application, the utility can appeal the decision to a state regulatory body, typically the state public utility commission. In some states, these bodies may have statutory authority to overrule local government decisions on the siting of generation and transmission facilities.

The time needed to develop a transmission line depends on the siting and permitting process and the time needed for construction. Projects that cross several states and are subject to multiple state jurisdictions are more complicated and take more time. In these situations, the actual construction can take less time than siting and permitting. Proposed new transmission projects can become controversial if project opponents believe the transmission project will have negative impacts on electric rates, the environment, property rights and values, or state and federal land. Disagreements may also occur over whether the utility has significantly and fairly evaluated potential alternatives.

For small transmission projects or upgrades of existing transmission projects, regulatory approvals can usually be obtained in a year or less. For long-distance transmission projects, however, the approval process is longer, although the amount of time varies for each proposed transmission line. For example, the American Transmission Company's 220-mile transmission line from Duluth, Minnesota, to Wausau, Wisconsin, took eight years to be approved but only two years to build, and American Electric Power's 765-kV, 90-mile transmission line from West Virginia to Virginia took 14 years to receive regulatory approval. In contrast, wind projects typically take one to three years to be approved and built. Lack of transmission can be a significant impediment to development of new wind projects; companies developing wind projects in areas lacking sufficient transmission often have no choice but to wait until new transmission is available (Daryanian et al. 2009).

Interconnection

Generator Interconnection. Planners should also be aware that interconnecting new generation facilities to the electric grid is another important factor for large wind energy projects. FERC has jurisdiction over entities that own, control, or operate interstate transmission facilities, while most distributed generation facilities are governed by state laws and procedures.

The majority of large generation (greater than 20 MW) interconnections in the continental United States are ultimately subject to FERC jurisdiction and approval, though the actual transmission studies are typically performed by transmission owners. Three studies are required in the interconnection process, each involving more detail and financial commitment than the previous one. First, a feasibility study provides a high-level look at a project's chosen point of entry into the grid. Next, an interconnection system impact study goes into greater detail and evaluates the impact of the proposed project on system reliability. Finally, an interconnection facilities study identifies required equipment, engineering, procurement, and construction work, and it estimates costs and timeframe of any necessary grid upgrades or improvements.

The vast majority of small residential and commercial systems (those having a capacity of 20 MW or less) interconnect with utility-level distribution systems and are therefore under the jurisdiction of state-level regulations. State-level interconnection procedures for non-FERC jurisdictional projects have set various size limits ranging from fewer than 100 kW to 80

(continued from page 64)

local, state, and federal permitting agencies; Native American tribes; and environmental and public interest organizations. Finally, the work of the group was open to review by all participants and any other interested parties.

Another important feature of RETI was its equal focus on environmental and economic factors. As some of the lengthiest battles over utility and transmission infrastructure development are fought over environmental impacts, bringing those considerations to the forefront of the planning process is vital. The RETI group developed a methodology that quantified both economic factors and environmental impacts and analyzed more than 30 CREZ areas on a "bubble chart" with *x*- and *y*-axes of environmental score and economic ranking score, respectively (Figure 4.1). The CREZs in the lower left quadrant of the chart had both the fewest environmental concerns and the lowest costs and highest economic values per unit energy production; they therefore emerged as priority areas for transmission development.

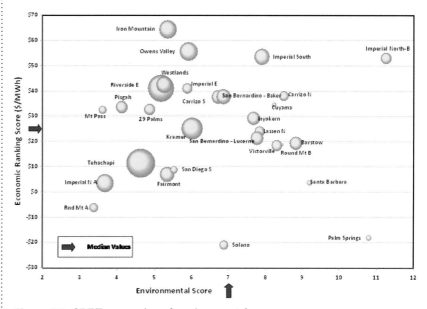

Figure 4.1. *CREZ economic and environmental scores*

Source: RETI Stakeholder Steering Committee

Notes:

Areas of the bubbles are proportional to CREZ energy.

Lassen South CREZ is off the right side of the chart. (Economic Score = 18, Environmental Score = 19.50, Energy = 1051 GWh)

San Diego North Central CREZ is off the right side of the chart. (Economic Score = 15, Environmental Score = 22.3, Energy = 502 GWh)

Victorville and Round Mountain-B are coincident

The RETI process was thus able to produce plans that could generate support from both the energy and environmental sectors. It also raised the level of awareness about the importance of environmental issues among utility planners and transmission engineers, groups that typically do not consider these issues in their technical planning. The analyses produced by RETI are now being used to inform the next level of transmission planning and implementation carried out by groups such as the California Transmission Planning Group, the California ISO, and the California Public Utilities Commission.

For more information, see www.energy.ca.gov/reti/index.html.

MW facilities, while some procedures contain no specified size limits and can therefore be used to interconnect any project as long as the interconnection does not fall under FERC jurisdiction. Table 4.2 outlines the various state procedure size limitations. Facilities sized 20 MW or smaller do not often interconnect to transmission lines that fall under FERC jurisdiction, but because connecting generation facilities in the 5 to 20 MW range to the distribution system may be difficult, some generation facilities in this capacity range do connect to FERC-regulated transmission lines. As of July 2010, all but eight states had created some type of state interconnection procedures.

TABLE 4.2. STATE INTERCONNECTION PROCEDURE SIZE LIMITATIONS

Size Limit (MW)	States
none	California, Hawaii, Indiana, Kentucky, Maine, Massachusetts, Michigan, New Jersey, North Carolina, Pennsylvania, Vermont
80	Iowa, New Mexico
20	Connecticut, Nevada, Ohio, Virginia, Washington
15	Wisconsin
10	Arizona*, Colorado, District of Columbia, Illinois, Maryland, Minnesota, South Dakota, Texas
2	Florida, New York, Oregon**, West Virginia, Utah**
1	Delaware
Under 500 kW	Arkansas, Georgia, Kansas, Louisiana, Missouri, Montana, Nebraska, New Hampshire, South Carolina, Wyoming
No state procedures	Alabama, Alaska, Idaho, Oklahoma, Mississippi, North Dakota, Rhode Island, Tennessee

Source: Fink, Porter, and Rogers 2010

*Arizona Corporate Commission procedures are still voluntary and recommended only. The utilities have implemented various size limits.

**Oregon and Utah impose a separate 25 kW size limit on residential systems.

There is currently no standard-form state interconnection procedure; requirements vary greatly from state to state. State interconnection procedures often incorporate certain general characteristics, including applicability, standard agreements and application processes, expedited processing for smaller systems, interconnection costs, and insurance requirements. Some state procedures apply only to certain types of utilities or certain technologies, most often renewable energy distributed resources. Standard agreements can facilitate generator interconnection by ensuring that project applicants know exactly what to expect and what is required with respect to their applications. Simplified standard agreements for small distributed-generation systems, along with simplified technical screening that permits systems to interconnect without further studies as long as they meet certain requirements, are particularly important for small-business and residential customers, as they can be deterred by complex procedures and long legal documents. Most state interconnection procedures have created several levels of review and documentation based on system size, with simplified processes for smaller inverter-based systems.

Interconnection costs for distributed generators include related application and connection fees and engineering, technical, and equipment charges; state procedures generally outline which costs the customers are responsible for paying. Some states also require customers to carry additional liability insurance, but because liability insurance is usually included in most standard small business and home owner insurance policies, many states simply specify how much general liability insurance a project owner must carry, instead of requiring additional insurance for the distributed generation system.

Several entities have developed model interconnection rules to assist state regulators in creating interconnection procedures for small and distributed generation, including the Interstate Renewable Energy Council (IREC), a nonprofit organization that creates and promotes the adoption of uniform renewable energy standards. The IREC procedures were originally published in 2006 and updated in 2009; they incorporate the best approaches and features found in previous models. The model offers four levels of review, three of which are based on project size (IREC 2009). Most applicable to smaller wind energy systems are Level 1, a set of simplified screens for inverter-based systems with a capacity of 25 kW or less, and Level 2, a set of screens for systems with a capacity of 2 MW or less, including those below 25 kW that did not pass Level 1 screening.

Implications for Planners

Transmission and interconnection are vital to the development of new wind projects. New transmission will be needed in many cases for the successful development of new wind energy capacity. Planners should be aware of what authority their local government has over the siting of generation and transmission assets and how this relates to state siting authority. As noted, local governments may have authority to approve the siting of generation and transmission facilities in conjunction with state regulatory agencies, but in other cases state regulatory agencies may override local government agencies.

Planners should also seek to provide input to electric utilities as they plan for new transmission or distribution facilities. Planners can provide feedback on how potential facilities may or may not be consistent with current plans by requiring public notice and local hearings for transmission facilities proposed by the state, as well as by intervening and filing comments in state regulatory dockets regarding proposed transmission facilities. Planners should request that sponsors of proposed transmission facilities comprehensively evaluate alternatives, including nontransmission alternatives such as energy efficiency or undergrounding proposed transmission lines, although that is generally much more expensive than installing overhead transmission lines (Fink, Porter, and Rogers 2010).

CHAPTER 5

Considering Wind Energy
in the Planning Process

Suzanne Rynne, AICP

 How can planners address wind energy in the community planning process? Planners have many opportunities in the planning process and in their day-to-day work to make a difference when it comes to wind energy. This chapter describes what planners do and where in the planning process opportunities exist to effect change. Since planners tend to deal with issues comprehensively, suggestions are provided both specifically with regard to wind as well as more broadly to renewable energy. Planners should think about wind energy within the broader scheme of renewable energy, energy policy, climate goals, and sustainability as it applies to their communities. These strategic points of intervention should also be communicated to those who may not be as familiar with the planning and community development processes.

STRATEGIC POINTS OF INTERVENTION

Long-Range Community Visioning and Goal Setting

Planners often conduct visioning exercises that produce long-term goals and objectives that community leaders look to when considering policies and actions. Community visioning is often the first step in developing a comprehensive, neighborhood, or downtown plan. Whether part of a planning process or on its own, visioning is an important first chance to identify new opportunities and priorities related to wind energy.

Here are some ideas for how planners can integrate wind energy into the visioning process:

- *Survey citizen attitudes.* Gauge the level of awareness and importance of wind energy to community members. In a community survey, for instance, ask questions such as, Are you in favor of renewable energy generation in the community?

- *Hold community workshops.* Consider how wind energy can be addressed through interactive forums. For example, in community workshops, create an exercise to gauge the level of support for renewable-energy options such as solar panels and wind turbines. Visualization techniques could also be utilized to show visual impacts of wind turbines in various locations.

- *Connect community goals.* Determine how wind energy is connected to other community goals and values. Review a list of these goals and values or a draft vision statement. For example, is sustainability or economic development part of the vision? Is your community concerned about climate change or interested in increasing local energy production? What about agricultural land preservation? How does wind energy fit into these goals or help your community achieve these goals? Discuss these connections with community members.

Communities may also use a visioning process to discuss new or existing goals and targets. For example, if a community or state has set a target for GHG emission reduction or a renewable-energy portfolio standard, the visioning process can be a good venue for discussing ideas about how to start meeting those goals or interim targets and figuring out what role wind energy plays in that regard.

Plan Making

Planning departments prepare plans of all kinds. They recommend actions involving infrastructure and facilities, land-use patterns, open space, transportation options, housing choice and affordability, and much more. Examining comprehensive plans and other planning documents to see if renewable energy is addressed and integrated into them is an important step.

Assessment and Analysis. An initial step in almost any planning process is a baseline assessment and analysis of existing conditions and trends. Establishing the baseline for energy use is critical to being able to track and measure progress toward goals. A good baseline measure of a community's energy use will take account of not just the amount it uses but also the mix of renewable and nonrenewable sources. These assessments can be summarized and included in an overall plan, for easy reference and connection to goals and policies that respond to them.

Comprehensive Plans. The comprehensive plan is a guiding document for the future of an entire community. It establishes goals and priorities and lays out action steps for meeting those goals. Planners should consider including an energy element in the comprehensive plan, integrating these issues within other elements, or both. Devoting an element to these issues may provide focus and allow communities to more easily amend an existing comprehensive plan. However, planners should also consider how energy use relates to other issues and elements in the comprehensive plan, such as land use.

Some states provide guidance on preparing local comprehensive plans, and communities should refer to this guidance in considering how to address wind energy in their plans.

Area Plans. In addition to a comprehensive plan, many communities also have more specific area plans—such as neighborhood plans, downtown plans, redevelopment district plans, and corridor plans—that might also incorporate wind energy. Planners should consider whether there are suitable locations for wind energy development and what policies might help it fit into other plans.

Functional Plans. Functional plans focus on a single issue or set of issues—such as transportation or open space—rather than a geographic area. Some functional plans are prepared by a municipality or a public or private special-purpose entity, such as a utility, an authority, or a school district. A community sustainability plan, climate action plan, or energy plan may provide an overview of energy use and sources (such as wind) in a community, as well as strategies for ensuring energy security in the future.

Standards, Policies, and Incentives

Planners write and amend standards, policies, and incentives that have an important influence on what, where, and how things get built and what, where, and how land and buildings are preserved. When updating regulations, planners should consider how zoning codes, building codes, subdivision codes, and other regulations and ordinances address wind energy.

McLEAN COUNTY, ILLINOIS

McLean County, Illinois, population 169,572, is a strongly agricultural county in the center of the state. It is home to two major wind farms: Twin Groves and White Oak. Twin Groves, developed in two phases, consists of 240 1.65-MW turbines, for an installed capacity of 198 MW in each phase. White Oak includes 100 1.5-MW turbines, with a total installed capacity of 150 MW. Combined, the installed capacity at these two wind farms is enough to power approximately 153,000 homes.

McLean County initially became interested in wind energy in 2004 when it was approached by wind farm developers. Since then, it has amended the zoning ordinance to allow commercial wind energy as a special permit major-utility use, subject to certain standards. These include a 2,000-foot distancing requirement from residential districts, height limits of 499 feet (or 200 feet if within 1.5 miles of a municipality with a population of 25,000 or more), and other safety and aesthetic requirements. According to Philip Dick, director of building and zoning, the flexibility of the special use process has contributed to the success of wind energy development in the county. He adds that by not adopting exhaustive wind energy regulations, the County can employ the special use process to explore the possibility of wind farms while still requiring developers to prepare necessary studies and meet basic conditions applicable to all special uses.

Used with permission of the McLean County Department of Building and Zoning

Another factor contributing to successful implementation of the wind farms is the county's—and community's—acceptance of wind energy development as a compatible use in agricultural zoning districts; wind energy is just another resource to harvest. Some objections have come from nonfarm residents in rural areas, but because the county limits residential development in agricultural areas this is not a large source of opposition. The County also uses the public hearing process to address individual concerns such as noise and drainage on access roads, in response to which developers have relocated proposed turbines to help mitigate potential impacts. Landowners also benefit from lease payments for the turbines on their properties, and the County benefits from the increased property tax revenues, a portion of which goes to local schools.

With interest in wind energy continuing to grow, county officials updated the zoning ordinance to address small wind energy systems as well, using a model wind energy ordinance written for Illinois as guidance. Small wind systems are now a permitted use in all zoning districts subject to standards, including setbacks of 110 percent of turbine height from rights-of-way and property lines, noise limits of 60 decibels at closest property lines, and height limits ranging from 50 to 150 feet based on parcel size. If the proposed site is in a crop-dusting agricultural area, applicants must also notify crop-dusting businesses prior to submitting their building permit applications. Additionally, the most recent update to the McLean County Regional Comprehensive Plan references the area's wind energy resource and recent wind energy development, noting that wind energy development can complement the county's agricultural heritage while offering significant tax dollars to the County as well as rental income to individual farmers.

(continued on page 72)

(continued from page 71)

For more information:

- Model Ordinance Regulating the Siting of Wind Energy Conversion Systems in Illinois (2003). Available at www.illinoiswind.org/resources/pdf/WindOrdinace.pdf.

- McLean County Regional Comprehensive Plan: A Guide to Sensible Growth Through Regional Cooperation (2009). Available at www.mcplan.org/community/regional_plan/REG_PLAN.pdf.

- McLean County Zoning Ordinance (2010). Article 6, Section 602.41, Use Standards—Utility, Major; Section 602.50, Use Standards—Small Wind Energy System. Available at www.mcleancountyil.gov/DocumentView.aspx?DID=45.

Development Codes

Zoning Code. Perhaps the most important regulatory tool for development in a community, the zoning code typically establishes permitted uses in various locations and provides standards for intensity of use, such as lot size, floor area ratio, setbacks, building heights, and permitted accessory structures. Planners should develop standards that support appropriate wind energy development and remove unintended barriers to it, such as by updating height restrictions.

Wind Energy Ordinance. Many communities opt to use a stand-alone ordinance or separate chapter of the municipal code to enact standards for wind turbines. Planners should ensure that the ordinance does not conflict with the zoning regulations and clearly states which regulation has primacy should a conflict be discovered in the future.

Incentives. There are a number of incentives available from federal and state sources, as well as utilities, that promote energy efficiency and the use of renewable energy. The Database of State Incentives for Renewable Energy (DSIRE), for example, provides links to information on many incentive programs that promote renewable energy and energy efficiency (www.dsireusa.org). Planners should consider creating a fact sheet with information on available incentives for developers and residents, as these incentives can be helpful in meeting energy goals.

Local governments also can create their own incentives. These may include expedited plan review for projects that meet or exceed established objectives; a waiver of permit fees, rebates, or other financial incentives to developers whose projects meet predetermined standards; and provision of technical assistance to help developers meet new goals and standards.

Development Work. Planners play an important role in development in their communities. They review project applications for consistency with applicable plans and regulations and may be involved in public-private partnerships to develop new projects. In reviewing private development projects, planners assess whether standards in ordinances and regulations have been met. Thus, a checklist of energy standards or goals for new projects can be helpful. If the goals are not mandatory, an expedited plan review and permit-processing track for projects that meet or exceed those goals can be an effective tool.

Public Investment

Cities and counties undertake major investments in infrastructure and community facilities. Planners should consider whether wind energy plays a role here also—for example, a wind turbine might provide power to a local school while also serving as an educational tool. Similarly, installing a wind turbine at a public facility such as a sewer plant might reduce energy cost uncertainty for the municipality over the long term.

Education and Outreach

The importance of education and outreach in implementing wind energy should not be overlooked. Community education and outreach activities often happen within each of the points of intervention highlighted in this chapter. Communities also hold separate education and outreach programs for important issues. In general, planners should consider ways to engage the public in discussing wind energy and provide educational forums for citizens to learn about wind energy. Public hearings should also be held when major wind projects are in development, to allow community members to express their concerns. Holding these hearings early in the process as well as throughout the process can help in understanding and addressing community concerns. (See the case studies for examples.)

In addition, planners should consider how to reach out to other agencies and stakeholders that influence and affect wind energy development. These might be neighboring jurisdictions, school districts, transportation agencies, and local utilities. Involve them early in local planning processes, get their input and feedback on new policies, regulations, and developments, and work with them to implement new standards. Furthermore, participate as these agencies create their own policy documents to ensure that wind energy and other local planning concerns are addressed.

CHAPTER 6

Regulating Small-Scale Wind
Energy Systems at the Local Level

Erica Heller, AICP

 Small wind energy conversion systems (WECS) are defined in the wind industry as those with a rated output capacity of 100 kW of less. Unfortunately, this definition says little about the size or characteristics of small WECS. Local planners more typically consider small WECS as those that are used to provide on-site power. This indicates both that small WECS are an accessory use and begins to establish a framework that ties the size limitations of small WECS to the power needs of the home, farm, or business that is the primary user.

Small WECS come in a wide range of sizes and styles. Figure 6.1 depicts a range of traditional "fan" or horizontal-style small WECS. Figure 6.2 shows another style of small WECS.

Small WECS technology is changing rapidly as new products are developed to enter this emerging market. While this dynamism can create new possibilities for how and where wind turbines are placed, it can also make writing small WECS regulations more challenging, as planners try to address a range of existing and potential configurations.

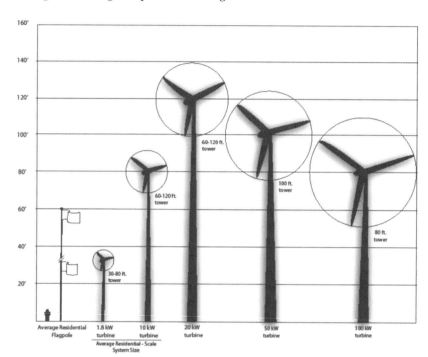

Figure 6.1. *Different kinds of fan-style small WECS*

Source: AWEA

Figure 6.2. *A vertical-style small WECS*

Source: AWEA

CHECKLIST FOR ORDINANCES

One recurring theme in successful wind energy implementation is the importance of having a local wind energy ordinance in place to clarify and streamline the WECS approval process. Not having an ordinance in place can delay project reviews. There are a number of model wind energy ordinances available, written by various organizations and states to serve as guidelines for communities. With the right ordinance language, planners can allow and encourage context-appropriate wind energy development while setting standards that protect communities from unwanted impacts.

Though every community should create a wind energy ordinance appropriate to its specific context and policy directives, there are common elements to such ordinances. The following ordinance framework is based upon a review of model wind energy ordinances and wind energy system permitting guidebooks.

A typical small wind energy system ordinance contains the following elements:

- *Definitions.* These typically specify that the system is intended to produce energy for on-site consumption, though the ordinance should allow for the possibility of reverse or net-metering, where electricity produced by the system that exceeds demand may be sold back to the utility or credited to the property's utility bill. As noted above, small systems are often defined by the maximum-rated capacity of the turbines as well; kilowatt thresholds used by various ordinances range from 5 to 100 kW.

- *Allowed use.* Many ordinances allow small wind energy systems by right in all districts as accessory uses, as long as development standards are met. For systems that may not meet all requirements, communities may allow approval through the conditional or special use process.

- *Setbacks.* In order to reduce noise and visual impacts and to minimize the slight risk of a possible tower collapse, communities require that freestanding turbines be set back from structures, property lines, or public rights-of-way a certain distance, typically equal to the height of the tower.

- *Tower height.* Some ordinances limit small wind energy systems to specific heights—for example, anywhere from 65 to 150 feet—while other ordinances exempt wind energy systems from district limits or allow them to exceed those limits by a specified amount. In some cases, turbine height is controlled indirectly through setback requirements. Placing absolute height limits on towers can sometimes prevent full utilization of the wind resource.

- *Visual appearance.* Ordinances typically specify that turbines not be lighted unless required by the Federal Aviation Administration (FAA)—though small residential turbines will usually fall under the 200-foot threshold above which FAA lighting is required—and that any signage on them other than turbine manufacturer or owner identification be prohibited. Some also specify that turbines be an off-white or gray color with a nonreflective finish.

- *Sound.* In order to avoid potential noise impacts for neighbors, ordinances typically require that audible sound from the turbine at the property line not exceed maximum noise limits established elsewhere in the zoning code. Some ordinances set specific noise thresholds, often 40 dBA to 55 dBA. Some allow exceptions for short-term events like storms when ambient noise increases.

- *Safety.* For safety reasons, ordinances typically require turbines to be designed and secured so as to prevent unauthorized climbing—for example, by prohibiting climbing aids from the first eight or 12 feet of the turbine pole.

- *Approved wind turbine design.* Some ordinances require that turbine be of a model approved by state standards or a recognized certification program. With the rapid proliferation of small WECS technology, certification provides a reassurance that the selected turbine has a safe and effective design.

- *Abandonment.* If the turbine is inoperable for a certain period of time, typically six months to one year, most ordinances require the owner to remove it to prevent a potential nuisance or safety hazard.

- *Permitting process and requirements.* See below.

Other elements that appear in some ordinances but less frequently than the elements above include the following:

- *Number per lot.* Some ordinances restrict small wind energy systems to one per lot. While this might be appropriate in dense urban environments, this may prohibit maximum use of wind resources on larger or more rural residential properties.

- *Minimum lot size.* Some ordinances restrict small wind energy systems to lots over a certain size threshold, often one acre. However, setting such a threshold may limit residents' full utilization of wind energy technology where noise and setback standards adequately protect neighbors from nuisance impacts; setback requirements also indirectly act to limit turbine placement on small lots.

- *Blade clearance.* Some ordinances specify that turbine blades must come no closer to the ground than a certain distance—for example, 30 feet—and also set minimum distances from the blades to structures and trees.

- *Automatic overspeed controls.* Some ordinances require that turbines be equipped with an automatic shutoff to prevent turbines from spinning too quickly in high-wind-speed or storm conditions.

- *Electromagnetic interference avoidance.* Although this is primarily a concern with large WECS, some small WECS ordinances specify that turbines must not cause radio or television interference and must be modified to eliminate any interference if they are found to do so.

- *Compliance with laws and regulations.* Some ordinances include a blanket statement that proposed turbine construction and operation must conform with local, state, and federal requirements, including building codes, electrical codes, and FAA requirements.

- *Maintenance.* Some ordinances specify that turbines must be maintained in basic working order.

- *Insurance.* Some ordinances require owners of small wind energy systems to hold liability insurance of a certain amount, though typically small wind systems can be added to existing home-owner policies as an appurtenant structure.

PERMIT PROCESSING

Because the impacts of small, residential-scale wind turbines are often minimal, local permitting processes for these uses can be streamlined and simplified to reduce barriers to their implementation. As a by-right accessory use, small wind energy systems may be handled administratively via the building or zoning permit application process through which compliance with the development standards laid out in the ordinance is demonstrated.

Ordinances should also require documentation to ensure safe turbine installation and operation, such as the following:

- a written description of the turbine;

- a site plan showing the planned location of the turbine on the parcel and its relation to other structures, property lines, public rights-of-way, and overhead utility lines;

- manufacturer- or engineer-certified drawings or plans demonstrating compliance with the Uniform Building Code;

- electrical component drawings, typically provided by the manufacturer, demonstrating compliance with the National Electric Code;

- evidence that the utility has been informed of the intent to interconnect the turbine to the grid, in cases where grid interconnection will occur; and

- for building-mounted small wind energy systems, a certified structural analysis of the roof or wall.

In general, planners should ensure that permit requirements are as simple and straightforward as possible, to minimize obstacles to home owners and commercial developers who wish to implement small-scale wind energy on their properties. The development standards laid out in a good ordinance for small wind energy systems will ensure that these by-right installations will have no nuisance impacts on neighboring properties, and the conditional or special use process should effectively handle turbine applications with the potential for more significant impacts.

ELEMENTS OF A SMALL WIND ENERGY ORDINANCE

WECS regulations must address potential impacts on neighboring properties, including location, height, aesthetics, sound, and safety. This section presents regulations that localities may use to address such impacts. When writing WECS regulations, planners should consider what broadly similar uses are regulated in the community and write similar regulations for WECS.

Districts and Uses

The first consideration of many planners drafting WECS ordinances is what districts they may fit into and whether the use should be processed as a permitted by-right use or as a conditional use.

Zoning Districts. Every WECS ordinance must address the zoning districts in which small WECS are allowed. In the past, many communities' first WECS permit applications came from residential property owners, and some communities' ordinances addressed only residential and agricultural zoning districts as locations for WECS. However, nonresidential zoning districts can provide great locations for small WECS. They are often less controversial, as potential impacts are more similar to other nonresidential uses. The noise of a small WECS for example, may be indistinguishable from the other uses and surrounding features—such as highways—in a commercial or industrial district. In addition, an increasing number of communities are finding that the first WECS applicants are commercial property owners. In some cases, they are WECS installers who want an on-site demonstration model. In others, a local business may either be interested in attracting positive attention or looking to offset energy costs. Thus, localities crafting ordinances should address a wide range of possible zoning districts as locations for small WECS.

Accessory Use Subject to Standards. The purpose of a small WECS is typically to provide power to a primary on-site use. As such, small WECS are accessory uses and should be designated as such. In agricultural areas the primary use (e.g., raising crops) may not be associated with a structure, but nonetheless small WECS are ancillary to it.

There is ongoing debate about whether wind energy development, particularly at the residential scale, should be allowed by-right according to the zoning ordinance or only as a special use, requiring more careful review and issuance of a special use permit. Small WECS potential impacts are

STAND-ALONE OR INTEGRATED ORDINANCE?

Both stand-alone WECS ordinances and standards that are integrated into existing zoning codes can be effective regulations for small WECS. Overall, the content of either is more important than the form. However, there are several advantages to integrating WECS regulations into an existing zoning ordinance. First, this location is often easier for applicants to find and access than a stand-alone ordinance. Second, applicants often find it easier to understand where and how the use is allowed when small WECS are clearly defined as accessory uses and integrated into the existing structure of use regulations. Finally, it is often easier for planners, who must review the permits, to administer an integrated regulation than a stand-alone ordinance.

reasonably predictable and it is very possible to draft standards to address them. Therefore, many small wind experts recommend that small WECS be designated as by-right uses subject to performance standards, rather than as conditional uses.

Designation as a by-right use subject to standards provides substantial reassurance to applicants and installers that each small WECS application will receive fair and consistent treatment, and discourages would-be NIM-BYs. If a WECS fails to meet the performance standards (such as maximum noise level) after installation, the issued permit may be considered null and the community can immediately take enforcement action. To provide regulatory flexibility, a proposed small WECS that does not meet the performance measures of the use standards may be reviewed as a conditional use.

The American Wind Energy Association (AWEA) advocates that small wind energy development should be a permitted use, requiring only a permit that allows a small wind system by default, provided it meets applicable design standards, such as those applied commonly to flagpoles, church steeples, and grain silos (AWEA 2008b). AWEA advocates against treatment of small wind energy development as a special or conditional use, for which installation is allowed under certain conditions identified in the statute. This option usually requires a more detailed description of the project from the applicant and often involves a public hearing. A hearing can disadvantage both the zoning board and potential owner, however, since each application must be decided on a case-by-case basis and educational efforts must begin from scratch. This process can cost potential small-turbine owners thousands of dollars and take hundreds of hours to accomplish (AWEA 2008b).

Table 8.1 (page 111) indicates which case-study communities in this report have allowed wind energy development by right versus through the special use process and at which scales. The notes to the table provide detail about case-study communities that have made exceptions to the by right–versus–special use dichotomy. In line with the recommendation of AWEA, most case-study communities allow small wind energy systems by-right and do not require a special use permit. Every location is different, and planners need community stakeholder input to successfully incorporate wind energy development into any local land-use regulations.

A few communities, such as Anchorage, Alaska, have made notification and approval by neighbors—whether a simple majority or larger percentage—a criteria for administrative permit approval. If the local community has an established precedent of requiring neighbor approval for ancillary uses, such a provision may be reasonable for WECS. However, this approach also has the potential to pit neighbors against each other and invite a fight over every small WECS application. Therefore, and because small WECS impacts can readily be regulated through standards, many small wind industry experts do not recommend this approach. In contrast, administrative review reduces permit review times, fees, and public resources needed for hearings.

Location and Setbacks

Most small WECS regulations include standards to address where a small WECS may be placed within a lot. The most common location standard is a setback standard. As noted, setback standards equal to the total height of the turbine are the most common, while some add an additional factor such as 1.1 times the total height. "Total height" is defined as maximum height of the turbine measured with the rotor blade(s) in the tallest operational position. The setback most often is measured from property lines, but in more rural jurisdictions may be established from others' homes. Setbacks may also be established from roadways, power lines, rail lines, and other conveyances.

SAN BERNARDINO COUNTY, CALIFORNIA

San Bernardino County, California, has permitted more small wind energy systems and has been regulating this use for longer than almost any community in the nation. This large Southern California county has a total population of more than two million people, 296,285 of whom live in unincorporated areas.

The county first adopted standards for small wind energy conversion systems (WECS) in 2002. Prior to that, it regulated them through an exception to the height limitations of the code. The county has revised its WECS regulations twice since 2002, most recently in 2010. In recent years, it has typically processed several small WECS permits each month, and as of February 2011 had issued more than 230 permits for small WECS.

All California local governments are subject to state law AB1207, adopted in 2001, which establishes procedures and standards for permitting small WECS. The intent of the law is to prohibit local governments from enacting regulations that would limit property owners from generating alternative energy for their own uses. Local government WECS regulations may be less restrictive but not more restrictive than the guidance in state law.

San Bernardino County has addressed wind energy in local land-use documents as well as through incentives. The general plan sets goals that encourage the use of alternative energy resources, establishing a policy basis of strong support for alternative energy tempered by concerns about environmental impacts such as wildlife impacts and water use. According to local officials, this support has led the planning commission to approve most WECS applications. As part of policy implementation for energy efficiency, the county has established a program to waive building permit fees on alternative energy technologies, including WECS. The money is allocated on a first-come, first-served basis each fiscal year, up to a maximum of $5,000 per project (significantly more than the cost of a typical small WECS permit). Since its inception in 2007 through June 2010, the county had waived $150,987 in fees. The 2010–2011 fiscal year budget allocation for the program is $45,000.

San Bernardino County does not define the size of WECS based on rated output. The county treats all "accessory" WECS—defined as those that generate energy for on-site use—as by-right uses subject to standards. WECS that generate power intended for off-site users are regulated as renewable energy facilities through the conditional use permit process.

San Bernardino County's accessory WECS standards allow a baseline of one WECS per parcel, though in 2010 the ordinance was amended to allow more turbines; on larger parcels, one WECS for every 10 acres up to a maximum of three is allowable, and for turbines under 50 feet in height, two per five acres are allowed, with an additional turbine allowed for every additional five acres up to a maximum of five. In residential districts, WECS are limited in height to 52.5 feet, though elsewhere allowed heights range from 65 to 120 feet, depending on the zoning district and region (valley, mountain, or desert). Setbacks were originally set at 125

percent of turbine height but in 2007 were reduced to a distance equal to turbine height. Compliance with noise performance standards is required.

The ordinance goes into more detail around visual effects and location: turbines may not obstruct views of adjacent property owners, must be placed below ridgelines when viewed from designated scenic corridors, and may not be located in scenic corridors or on designated historic properties. The 2002 ordinance included a provision that WECS must be "earthtone" in color; this was amended to "nonreflective, non-obtrusive" in 2007, and the entire provision was deleted in the 2010 revisions after representatives of the Bergey Windpower company—whose turbines are a signature bright-yellow color—suggested that the provision unfairly affected their company. In addition, the permitting process was changed to administrative review from the original building permit requirement in 2007, and a neighbor-notification provision was added.

San Bernardino Land Use Services Department

In 2010, the county added a chapter to its zoning ordinance for large renewable energy–generation facilities, including specific standards for wind energy development. Renewable energy facilities are permitted as conditional uses in resource conservation, agriculture, floodway, rural living, industrial, and institutional districts. The ordinance imposes turbine height limits of 500 feet and a number of setback requirements, including the lesser of two times turbine height or 500 feet to exterior project boundaries; the greater of 1.5 times turbine height or 1,500 feet from off-site residences; one times turbine height from on-site residences; and 1.5 times turbine height from streets, rail, trails, or public access easements. The ordinance also requires FAA compliance and clearance from the Department of Defense and provides detailed decommissioning requirements.

The County has revised its regulations over time in response to both changes in the state law and local experience. Height limits balance the desires of local installers and owners who push for higher limits against the concerns of other property owners who want to reduce aesthetic impacts. Initially, the county required a simple building permit for WECS, but this was changed to administrative review after an installed turbine that had been approved

(continued on page 82)

(continued from page 81)

through a building permit on a property in a hilly neighborhood stretched up almost directly in front of the picture window of a neighbor's house up the hill, blocking his views. This led to consensus in the county that administrative site plan review was needed to approve the location of the WECS. The county chose administrative review, rather than a conditional use permit as allowed by state law, to speed permit decision times and keep permit costs low.

San Bernardino County continues to adapt its small WECS regulations to respond to evolving technology. As of early 2011, the county was in the process of considering additional revisions to better address "microturbines." The intent, according to Jim Squire, assistant director of the Land Use Services Department, is to adapt the standards to address the smallest WECS, as they appear to fit into more urbanized areas with few potential impacts on neighbors.

For more information:

- County of San Bernadino 2007 General Plan. Section V, Part 7, Energy. Available at www.sbcounty.gov/ehlus/Depts/Planning/documents/FINALGeneralPlanText3-1-07_w_Images.pdf.

- County of San Bernadino 2007 Development Code (amended 2011). Division 4, Chapter 85.18, Accessory Wind Energy System Permit; Chapter 84.26, Wind Energy Systems—Accessory; Chapter 84.29, Section 84.29.030, Renewable Energy Generation Facilities: Wind Energy Development Standards. Available at www.sbcounty.gov/ehlus/Depts/Planning/documents/developmentCode.pdf. ◀

Other location standards may be used to add specificity about siting in relationship to land features and views. Because WECS need to be sited for optimal wind access, prohibiting them from ridges, bluffs, shorelines, and other visible areas is not recommended. However, some communities do restrict WECS within specific highly scenic areas, especially when such areas have been designated as being of particular importance in policy and planning documents. For example, the State of Vermont does not allow localities to restrict small WECS based on aesthetics except from designated historic areas and scenic byways. Some communities have enacted protections for important viewsheds of unique local features, such as from the civic center to a mountaintop. Restrictions from visual encroachments in these areas also apply to WECS. The San Bernardino County, California, WECS ordinance was amended to require site plan review of small WECS after one turbine in a hilly neighborhood ended up in front of the picture window of an uphill neighbor. The county adopted a more discretionary approach to approvals, which had been issued through building permits alone, and now requires site plan checks to review visual impact. (See sidebar.)

Height

Height standards for small WECS are often the most controversial regulations that local planners address. WECS must be placed higher than surrounding obstructions to allow them to function properly. However, controversy arises over how height may increase visual impact of the WECS and how much visual impact is acceptable to the neighbors and the community at large.

It is not uncommon for small WECS ordinances to limit turbine height to the limit for other structures in the zoning district. However, this may limit the ability of WECS to produce energy. WECS are highly sensitive to relatively small variations in wind speed and turbulence, which can be significantly affected by height and location within a parcel. (See Chapter 1.) The minimum clearance required for small WECS to function properly is 25 to 35 vertical feet above all surrounding obstructions that are within 300 horizontal feet. In the vast majority of locations and situations, applying zoning district height limits to small WECS will make them so inefficient as to be economically impractical. In essence, such a height restriction amounts to a de facto prohibition on small WECS in most residential and commercial zoning districts. In acknowledgment of this fact, several states, including California, Nevada, Wisconsin, New Jersey, and Michigan, have passed laws that limit the ability of local governments to prohibit WECS or adopt WECS regulations that substantially impede their functionality. One way, then, to address height is to exempt small WECS from zoning district height limitations similar to other common uses and projections, such as chimneys, flagpoles, radio antennas, power transmission towers, smokestacks, and wireless communications towers.

Some communities such as the City of Reno, Nevada, limit height only through setback requirements or through a combination of setback requirements and a generous maximum height. For example, the City of Alexandria, Minnesota, uses a 1.1-times setback as the height standard up to a maximum of 175 feet. On a small lot, such a setback requirement will dictate the maximum height of a small WECS. For example, a WECS located at the center of a 120-foot-square lot that meets this setback requirement can be no taller than 109 feet. In most cases, the centers of small lots are occupied by the primary structure, meaning the WECS must be placed closer to the property line and is thus subject to a lower maximum height in order to meet the setback requirement. Using the setback requirement as the effective height limitation is an elegant and practical way to tie the scale of allowed WECS to the density of development and to tailor it to the unique siting opportunities and limitations of a specific parcel.

Blade clearance from the ground is another height consideration, particularly on commercial or institutional properties where members of the public have access to the site. Such regulations apply to blades of any orientation. Common sense, protection of property, and liability concerns ensure that the vast majority of owners install WECS such that the lowest point of the spinning blades are well above any level that would make them easy to tamper with or that would interfere with the normal activities of persons or vehicles. Still, some localities choose to specify a minimum WECS blade clearance height, such as 25 feet.

Aesthetics

In addition to the visual impact posed by height, the aesthetics of a turbine may increase or reduce its visual impact or reactions to it. The color or style of a WECS and what signage may be posted on it are aesthetic concerns.

Some communities have sought to reduce visual impact by requiring that WECS blend into the surrounding environment, with mixed success. An early version of the WECS regulation in San Bernardino County, California, prescribed earth tones to blend with mountain or forest backgrounds. (See case study, page 81.) While the visual backdrop of a given WECS depends on the location and perspective of the viewer, from most perspectives WECS are framed against the sky, particularly if they are allowed to reach a height that clears nearby wind obstructions. Based on studies that show that white or matte gray colors blend in best against a range of skies, some small WECS ordinances (such as Rochester, Minnesota's) now require these colors. However, because some manufacturers of small WECS differentiate themselves in the marketplace with signature colors—bright yellow, in the case of the Bergey company—color regulation could be viewed as unfair to a particular manufacturer. Representatives of Bergey successfully encouraged San Bernardino County to add discretionary flexibility to its WECS color standard.

A typically noncontroversial and fairly common restriction is to prohibit commercial signage on WECS structures or blades. This is particularly common in residential areas, while some communities allow signage on WECS in commercial or industrial areas. Some communities restrict all commercial signage while expressly allowing electrical warning and other safety signs.

Some communities, out of aesthetic concerns, dictate what style of WECS may be allowed by district. For example, some urban communities—including Chicago—dictate that only rooftop-mounted WECS may be installed in residential districts. This is problematic for several reasons. First, technology is rapidly changing in the small WECS market, and specific style requirements are likely to become outdated. Also, the variations in the wind microclimate at a specific site or location may cause one style of WECS to be significantly more productive than another. This fact strongly suggests that qualified wind experts, rather than planners, should select the best WECS technology for an applicant's site. Prescriptions for roof-mounted WECS are particularly questionable, as a study of such systems in the United Kingdom showed that many of these units substantially underperformed manufacturers' projections (Encraft 2009). The Henderson, New York, WECS ordinance strikes a balance by allowing roof-mounted WECS in most zoning districts with a streamlined approval process, as well as free-standing small WECS in many of the zoning districts subject to additional standards and discretionary review.

Even communities that do not intentionally prescribe a certain style of small WECS may find their regulations are written in language that assumes fan-like rotor blades. These regulations can be challenging to apply when preparing and reviewing an application for a newer technology or configuration, such as vertical or helix-type WECS. The City of Reno, Nevada, which is home to one of the manufacturers of vertically oriented small WECS, has taken care to draft its small WECS regulations to make them applicable to a variety of styles of small WECS.

ROCKINGHAM COUNTY, VIRGINIA

Rockingham County, population 76,314, lies on Virginia's western border adjoining West Virginia, stretching from the Shenandoah Valley west to the Appalachian Mountains. Much of the western part of the county lies within the George Washington National Forest. Wind energy is not a new concept in this region. Farmers have long used small windmills to pump water on dairy farms; some of these windmills are still in use by Mennonites in the eastern valley.

In the early 2000s, wind energy developers started showing interest in the region, and by 2004 the County had developed and approved an ordinance limiting wind energy systems to fewer than 65 or 80 feet in height. Four or five small wind projects were permitted under this ordinance. However, pressure on the county to enact further regulation intensified when a group of landowners in the northwestern portion of the county approached the Dominion Power utility about possible development of a wind farm. Studies showed a high-wind resource corridor along Appalachian ridge tops in the western portion of the county, and federal subsidies and the U.S. Department of Energy's "20% Wind by 2030" initiative were driving additional development interest. The landowners were joined by Massachusetts-based wind energy developer Solaya in asking the county to develop regulations for large wind energy development.

The sudden interest was a wake-up call for Rockingham County officials. They looked to other Shenandoah Valley jurisdictions for lessons and found the experience of nearby Highland County especially instructive. A recent wind farm proposal there had triggered special use permit review because of tower height. But because the county did not have large wind energy development regulations in place, the proposal went through Virginia's State Corporation Commission review process, which limited local officials' control over the final outcome. Although Highland County has since adopted local regulations regarding large-scale wind, Rockingham County officials did not want to follow the same path; they decided to pursue large-scale wind energy regulation through proactive collaboration.

To tackle this issue, the Rockingham County Board of Supervisors convened a diverse working group comprising developers, local citizens, conservationists, county officials, utilities, and experts such as Dr. Jonathan Miles, director of the Virginia Center for Wind Energy at James Madison University. The working group spent over a year developing an ordinance to address both commercial- and residential-scale wind energy development. Citizens and conservation groups were eager to gain developers' perspectives and insights on financing, other projects they had done, and regulations in other states. Members of the working group felt it was better to hash out debate in committee rather than try to resolve differences or tackle tougher issues in public hearings.

The group wrestled with several concerns. To study the environmental impacts of wind energy, they visited a wind farm at Mount Storm, West Virginia, to examine how that project addressed potential effects on birds and endangered species; they then decided to include Endangered Species Act language within the final ordinance. Noise was a concern for many and was discussed at length, though the group felt it lacked the technical expertise and understanding to come to any conclusions. Recreational users of the national forest feared that wind farm development might alter the area and limit opportunities for bird watching, hunting, horseback riding, and other activities, but the group was able to agree on conditions to prevent such negative impacts. Developers and citizens disagreed on minimum setback requirements, necessitating in-depth examinations of the technical issues involved. Some participants felt that developers' greater representation within the group resulted in recommended setback language that favored the development perspective. However, others felt the total composition of the group ensured that no single interest was overrepresented. Despite differences among the participants, the process was congenial and provided opportunities for members to network and resolve issues.

The working group ultimately produced a draft ordinance, which was refined by the planning commission, board of supervisors, and the public through several hearings. Though some members of the group were more satisfied with the final product than others, all agreed it was better to keep regulation local instead of deferring that power to the Commonwealth of Virginia.

The new ordinance was adopted in November 2010. It retained provisions for small wind energy systems, limiting turbines to no more than 80 feet in height and 100 kW in output and allowing them by right with administrative review in certain zoning districts. The ordinance added new language to address large-scale wind energy systems, allowing them as special uses in agricultural and public service zoning districts, subject to local environmental as well as state and federal requirements. The ordinance sets turbine height limits of 500 feet and setbacks of 125 percent of structure height from nonparticipating property lines and the greater of 160 percent of structure height or 800 feet from structures or public use areas. The review process requires two public information hearings, one before application submission and the other prior to the special use permit public hearing. The ordinance also provides for project decommissioning and restoration of project sites.

Despite Rockingham County's efforts, there has been some criticism of both its approach and the end result. While County officials did pass an ordinance that regulated both small and large wind energy development, some critics—including John Hutchinson, AICP, of the Jennings Gap Partnership, who prepared a report on wind development in the Shenandoah Valley for the Rockingham Community Alliance—felt county officials and the working group sidestepped the larger issues of comprehensive planning for wind farms. Commercial-scale wind energy facilities have potential for significant impact on the valley, and such major new uses should require a more comprehensive zoning overhaul. However, even critics acknowledge that Rockingham County is better off with the new regulations than without them.

(continued on page 85)

Sound

Although noise is often one of the first concerns raised by neighbors and others wary of small WECS, it is readily addressed by newer technology and straightforward regulations. Most modern small turbines are quite quiet. The sound output of a 2 kW WECS (a size that is often appropriate to serve a single-family residence) is typically about 55 dBA at a distance of 50 feet away from the hub. This distance may be in any direction, including standing directly below a 50-foot tall WECS. Fifty-five dBA is about the level of a kitchen refrigerator—a level above which a conversation can easily be maintained.

Many communities' generally applicable noise regulations prohibit noise above 50 to 65 dBA from any use as measured at the property line. These regulations address small WECS noise adequately in standard conditions. However, in stormy conditions, sound output from small WECS may be higher, as is background noise such as wind in the trees. (Such weather events themselves typically reduce neighbors' tendency to engage in outdoor backyard activities where they would be annoyed by increases in WECS' sound output.)

Good locations for WECS may also be found in environments that are relatively noisy, such as industrial districts or near highways. In such environments, it may not make sense to strictly limit WECS to 55 dBA. In order to address the range of possible environments and events that reasonably accommodate more sound from WECS, many small WECS ordinances, including that of City of Hays, Kansas, include a context-sensitive sound standard such as "55 dBA or 10 decibels greater than ambient noise."

Safety

The safety of small WECS should be taken seriously by anyone drafting a WECS ordinance. Even so, it is important not to create fear-driven or overly burdensome requirements that do not reflect real risks. Since most small WECS are designed by certified electrical engineers, manufactured by reputable companies, and installed by competent building technicians, there is no more justification to require individual testing and certification of each unit and its installation than there would be for other small home machinery (e.g., furnaces). However, in the fast-growing market for small WECS, there have been some less reputable, lower-cost entrants into the market that may not be constructed with rigorous safety elements. The industry is responding with a voluntary certification system for small WECS. (See sidebar, page 87.)

It is appropriate to require inclusion of standard electrical drawings from the manufacturer and for footing inspection by competent building officials. While it is common for soil samples and wet stamps to be required for large WECS, small WECS structural loads are more similar to flagpoles and cell towers. In areas with standard soil conditions—not including gravel, sand, or muck—no soil study is needed for small WECS.

Many small WECS ordinances including AWEA's small wind model ordinance (available in AWEA 2008b) require small WECS to have manual override braking. This is a mechanical system that can stop WECS blades from spinning if wind conditions are so extreme that the electrical braking system fails.

Another safety concern, particularly in residential neighborhoods, is that small WECS might attract climbers. While some communities require small fenced areas around the base of WECS, these are difficult to maintain for the owner and often ineffective. A more effective and less expensive alternative is to remove or block climbing features below a reasonable height. On poles and many towers, climbing pegs or rungs should be removed below 12 feet. On lattice structures, sheets of solid metal or wood can be affixed to block climbing. Some industry experts suggest that a "Danger: High Voltage" sign may be an effective way to deter climbers.

(continued from page 84)

This may not be the final word on wind energy development in the county, however. According to local wind energy experts, the key to wind in western Virginia rests with the federal government because of its extensive public land holdings there. Most of the high ridges in the area are owned by the National Park Service or the Forest Service, and decisions by those agencies will supersede local government action on their lands. A forthcoming National Environmental Protection Act (NEPA) report on wind energy development within federally owned park and forest lands is expected to establish wind energy policy for public lands in the mountainous region along the Virginia–West Virginia border.

For more information:

- Rockingham County Code (2011). Chapter 17, Article XII, Division I, Small Wind Energy Systems; Division II, Large Wind Energy Systems. Available at http://library1.municode.com/default-now/home.htm?infobase=12196&doc_action=whatsnew.

- *Local Ordinances to Regulate Wind Energy Projects* (2009). Prepared by John Hutchinson, AICP, for the Shenandoah Valley Network and the Rockingham Community Alliance for Preservation. Available at www.fauquiercounty.gov/government/departments/BOS/pastagendas/05-14-09/Windmills_Att3.pdf.

CITY OF HAYS, KANSAS

The City of Hays, Kansas, is located in Ellis County; with a population of 20,510, it is the largest city in the northwestern part of the state. It lies just 60 miles west of the Smokey Hills wind farm in nearby Lincoln County. In 2009, a wind farm was proposed in Ellis County not far from Hays. The project became contentious due in part to a lack of county regulations. This motivated the city's superintendent of planning, inspection, and enforcement, Jesse Rohr, to pursue regulations for Hays before the city might face a similar application unprepared.

Hays has no policy guidance in place regarding alternative energy resources, though the city intends to address wind energy in the current comprehensive plan update. Though such guidance would have been helpful in crafting wind energy regulations, the planning staff drew from model ordinances, other communities' ordinances, and APA's publication *Zoning Practice* to craft their regulations, which were refined through community outreach and public hearings.

The resulting ordinance, adopted in 2010, allows wind energy conversion systems (WECS) as special uses within all districts, subject to strong but uncomplicated standards. It sets height limits—the most contentious issue in developing the ordinance—of 45 feet in residential districts and 125 feet in nonresidential districts and the city's three-mile buffer zone. The regulations require setbacks of 1.1 times total turbine height, limit noise to 55 dB or 10 dB above ambient conditions, restrict access by climbing, require submittal of electrical and engineering drawings as provided by the manufacturer, prohibit use of WECS as signage, and provide for removal of any abandoned WECS. Public opinion was generally in favor of small WECS—there is already one small turbine operating within the city limits—but not large turbines; accordingly, the 125-foot height restriction effectively prohibits utility-scale turbines and community wind projects from the area under the City's jurisdiction.

Soon after adoption of the ordinance, Fort Hays State University (FHSU), which is located in the three-mile buffer zone, began to investigate installing a larger WECS on its campus to offset utility costs. Studies showed that installing 10 125-foot turbines as required by height limitations was not a cost-effective alternative to two 500-foot turbines, so the university asked the City to consider a height exception for the project. City staff and officials generally supported the concept, but though most of the community supported a WECS taller than 125 feet on the FSHU site, they opposed taller WECS anywhere else in the City's jurisdiction. Officials could not find a legally defensible way to allow a height exception only for the FHSU site, so they rejected the proposal. FHSU has not since brought forward an application for the smaller turbines. When asked, Rohr agreed that it may be possible to establish policy in the comprehensive plan that would provide a sound basis for allowing 500-foot tall WECS at FHSU while maintaining the 125-foot limit in the remainder of Hays's buffer zone. Once the new plan is adopted, the community will make appropriate revisions to achieve consistency.

The process of developing a wind energy ordinance in Hays has not followed the ideal steps of setting broad land-use policy objectives before implementing detailed regulations. Rather, it is typical of the iterative, incremental process most communities must pursue for WECS and other emerging areas of land-use planning.

For more information:
- City of Hays Code of Ordinances (2011). Chapter 71, Article X, Wind Energy Conversion Systems. Available at http://library.municode.com/index.aspx?clientId=1423&stateId=16&stateName=Kansas. ◄

Many communities require that small WECS be removed if abandoned for a period of time. In the near term, removal requirements ensure that a community does not experience visual impacts from WECS that are not serving useful purposes. Such regulations also prevent potential hazards from unmaintained small WECS that might otherwise be left to fall into disrepair. Removal regulations typically use the community's standard established definition of an abandoned use. The Anchorage, Alaska, ordinance, for example, includes a removal requirement for small WECS if the use is discontinued for 12 months. However, if a community's standard definition of a discontinued use employs a short duration (e.g., three or six months) and local conditions limit WECS use to particular seasons, a longer duration should be allowed for small WECS than for other uses. The removal requirement and procedures should parallel those for similar abandoned structures or uses, such as billboards. Removal costs for discontinued small WECS are not great enough to warrant controversial bonding or escrow holdings as may be used for large wind farms.

Though safety considerations for large WECS related to interference with airspace and electronic communications rarely apply to small WECS, around airports and military installations it may be appropriate to consult such agencies, particularly for turbines larger than 10 kW. Similarly, special consideration and review may be warranted in close proximity to airports, helipads, or military facilities or where ground elevations increase rapidly under established flight paths, such as on hills or mountainsides within a mile of an air facility.

Wildlife Impacts
Small WECS have very limited wildlife impacts. Their blade areas simply do not create as much of a hazard as those of larger WECS, and there is typically plenty of maneuvering room around them. Ground disturbance is also minimal. The

number of birds killed annually by WECS is fewer than by housecats or glass windows and doors (Kerry & Curlinger n.d.). Even the Audubon Society has issued statements in support of well-located WECS (Audubon 2006). However, in areas with known endangered or critical species habitats where a community has significantly restricted other types of development, small WECS should also be considered for restrictions.

Quantity Standards

Some local regulations address the number of small WECS—both how many WECS may be allowed and what amount of energy may be generated. Early iterations of both types of standards were often unintentionally too strict. Better models are suggested below.

The number of WECS per lot is often a local concern and thus is addressed in many codes. Many first-generation WECS ordinances limit small WECS to one per lot. Though it is much more cost-effective to install one large small WECS than multiple smaller WECS, some owners might be inclined to install multiple WECS—such as WECS installers who want several as demonstrations or a heavy energy user. A maximum number of WECS per area of land more precisely addresses the impact concern of multiple WECS than a per lot limit. Lot size may vary greatly, particularly in industrial and agricultural areas. As for other standards, additional WECS per acre can be considered as conditional uses in unusual circumstances. The Municipality of Anchorage, Alaska, allows one WECS per lot subject to use standards in many zoning districts, while additional small WECS may be approved as a conditional use. (See page 88.)

Regulation of the quantity of WECS energy output relates to their definition as accessory uses. Net metering regulations are often very effective at limiting the size and energy production of a small WECS to a level that is well

Source: Kern County, California

CERTIFICATION FOR SMALL WECS

Several industry, governmental, and advocacy groups are working to establish a certification system for small WECS. Foremost among these appears to be the Small Wind Certification Council (SWCC; www.smallwindcertification.org).

In 2010, the SWCC began taking applications for certification. By April 2010, 21 manufacturers had applied for certification of 25 small WECS models. SWCC certifications will be issued only for models that meet the American Wind Energy Association's small WECS durability and safety standards, and they will result in labels that inform consumers about the model's rated energy output, power, and sound output. The certification process is somewhat lengthy (one to two years) in order to thoroughly test all the parameters, particularly real-world performance. Once certification is complete for a wide range of small WECS models, zoning ordinances may be written to require certified small WECS.

USE DATA TO COUNTER DRAMA

Carefully consider the source of anecdotes and videos that depict WECS falling over, breaking apart, catching on fire, or failing. Several video clips can easily be found online that show WECS failing. Many of them originally come from test facilities (which is why a camera was focused on the WECS at the moment of failure!). Such videos are not necessarily proof of the hazards of WECS. Rather, they may represent the responsible efforts of manufacturers to thoroughly test new WECS models and technologies prior to marketing them. There are very few documented instances of installed WECS failing due to mechanical, electrical, or structural failure.

CITY AND BOROUGH OF ANCHORAGE, ALASKA

The impetus for the City and Borough of Anchorage, Alaska, population 286,174, to adopt wind energy regulations was a common one: several property owners interested in installing turbines had approached the municipality about obtaining permits. The state had been promoting alternative energy, and several turbines had been installed in rural communities. Installers based in Anchorage asked the City to provide regulations to create certainty for potential customers.

In 2008, the municipality decided not to accept more permit applications until appropriate regulations could be adopted. Staff began researching and drafting an ordinance. The process took more than one year. The resulting ordinance allows WECS subject to standards in most zoning districts with administrative site plan review. Small WECS are limited to a rated output of 10 kW in residential districts and 25 kW in nonresidential districts. They are conditional uses in medium-to-high density residential districts or in cases where the applicant wants to install more than one WECS on a single property. The ordinance also allows utility-scale WECS as conditional uses in certain industrial and infrastructure districts. It requires analyses of wildlife impacts, visual impacts, and noise, as well as shadow flicker studies if the WECS will be located within 1,300 feet of habitable buildings; it provides for setbacks of three times WECS height from residential property lines.

The most challenging issues in crafting the small WECS ordinance proved to be concerns regarding visual impacts and turbines "looming" over neighbors' properties. To address these concerns, the Anchorage ordinance requires a visual impact analysis. The municipality also considered establishing a minimum lot size but eventually settled on a setback of 1.5 times total WECS height. Finally, the ordinance requires written consent from a simple majority of abutting residential property owners.

The ordinance was adopted in August 2010. There was no public opposition voiced to the ordinance at the adoption hearing, unusual for such a substantive issue. Review fees in Anchorage were reduced in early 2011, bringing the fee for small WECS administrative review from $3,300 down to about $1,600.

Since adopting the WECS ordinance, the City has received one proposal for a small WECS installation from Scott McKim, a teacher at Begich Middle School interested in installing a demonstration project through the Wind for Schools program of Wind Powering America and the National Renewable Energy Laboratory (www.windpoweringamerica .gov/schools_wfs_project.asp). Students learned about wind energy systems and, as part of the required solicitation for neighbor consent, hosted an open house to answer any concerns about WECS. They have not received any objections to the proposal. According to McKim, he has had a good experience working with the municipality and did not find the WECS regulations onerous, though the administrative fees are fairly high. (The Alaska administrators of the Wind for Schools program indicated that other cities had been willing to waive such fees for school demonstration projects.) In early 2011, the school received funding from the state to match the Wind for Schools grant, and the project will move forward.

For more information:

- Anchorage Municipal Code (2011). Title 21, Section 21.45.410, Small Wind Energy Conversion Systems; Section 21.50.470, Small Wind Energy Conversion Systems— Multiple Free-Standing Towers; Section 21.50.480, Utility Wind Energy Conversion Systems. Available at http://library.municode.com/index.aspx?clientID=12717& stateID=2&statename=Alaska.

matched to a property's primary use. However, such regulations vary significantly across states and utilities. Local regulations language should not be so specific or restrictive as to force owners to install a WECS on the basis of its maximum theoretical output in an ideal year rather than the realistic output in an average year. For example, an ordinance could say "Energy produced by an accessory WECS shall primarily serve the on-site use" rather than strictly limiting energy production to an amount not more than the demand of the primary use. Such language acknowledges variable annual conditions and production. In addition, it allows a property owner to occasionally provide excess energy to the grid, which has positive community and societal benefits.

Conclusion

Small WECS regulations need not be lengthy or especially complex to thoroughly address the potential impacts of this use. In fact, one of the most important considerations in drafting a small WECS ordinance may be to understand how the potential impacts differ from large WECS so as not to include unnecessary or onerous requirements for small WECS. Small WECS can be allowed in a wide range of zoning districts. They can be regulated effectively as by-right accessory uses subject to standards. Local standards must balance the visual impacts of small WECS against WECS functional need for wind access. Small WECS are a substantial investment on the part of a property owner. As for all uses, local government regulations for small WECS should be crafted to minimize the time and expense required for application completion review while protecting the community from negative impacts.

CHAPTER 7

Permitting Utility-Scale Wind Energy Systems at the Local Level

Kevin Rackstraw

Large, utility-scale wind energy systems have different impacts than small systems, and they require a completely different set of ordinance requirements to protect local communities from potential negative impacts of turbine siting and construction. Utility turbines are much larger, often several hundred feet tall, and wind farms usually involve multiple turbines, sometimes more than 100, spanning hundreds or thousands of acres. Roads must be constructed to access each turbine site, and the heavy construction equipment and large trucks used to bring turbine components to those sites can damage existing local roads. Electrical substations and grid interconnection infrastructure must also be constructed.

CHECKLIST FOR ORDINANCES

A community can best prepare to address such proposals through ordinance language specifically targeted at this use. Typical ordinances for large wind energy systems include the following elements:

- *Definitions.* Large wind energy systems are often defined as comprising one or more turbines for the purpose of generating electricity for commercial sale. Large systems are also often defined in terms of capacity—typically over 1 MW.

- *Allowed use.* Ordinances tend to require conditional or special use approvals for large wind energy systems, and they often limit this use to rural and industrial districts. Communities looking to encourage utility wind projects may make this use a by-right principal or accessory use in certain rural and industrial districts; some have created wind energy overlay districts to encourage the location of large wind projects in certain areas.

- *Setbacks.* As with small wind energy systems, ordinances often provide setback requirements in terms of the height of the turbine, often slightly more than the height of the turbine. Some ordinances also provide for setbacks of absolute distances, such as 1,000 feet from inhabited structures. Setbacks are typically required from structures, property lines, and public roads or rights-of-way. Ordinances may allow for setbacks to be decreased with signed agreements from landowners.

- *Tower height.* In many cases, large wind energy ordinances do not set height restrictions on commercial turbines, as turbines have tended to become ever taller as technology has evolved.

- *Electromagnetic interference.* Turbines must not cause microwave, television, radio, or navigation interference. Near military installations, radar interference should also be considered.

- *Visual appearance.* As is the case with small wind energy systems, many ordinances require that large turbines be of neutral color and nonreflective finish; that they be lighted per FAA guidelines with no additional lighting allowed; and that signage be limited to turbine manufacturer, facility owner or operator, and emergency contact information.

- *Noise.* Ordinances typically require audible turbine noise to be below specific sound thresholds at property lines, often 40 dBA to 55 dBA. More detailed noise standards may cast thresholds in terms of ambient noise levels.

- *Shadow flicker.* Shadow flicker can be an issue with large turbines. Its extent will change with the angle of the sun over the course of a year. Though turbines are usually sited to avoid shadow flicker on neighboring structures and the complaints that this may cause, some ordinances include provisions limiting shadow flicker on neighboring properties within a certain distance of turbines, often 2,500 feet, to a certain number of hours per year, often 30. Ordinances may allow both noise and shadow flicker restrictions to be waived with the signed consent of affected property owners.

- *Minimum ground clearance.* Some ordinances specify minimum ground clearances of turbine blades, usually between 12 and 50 feet.

- *Safety.* Large turbines must be designed to prevent unauthorized climbing; fencing of electrical substations and other utility structures is also required. Some ordinances require operators to post emergency contact information at the facility.

- *Decommissioning.* Ordinances for large wind energy systems require developers to decommission turbines if they are no longer being used. Ordinances may specify when decommissioning must be commenced and completed relative to the end of the turbine's useful life, as well as the degree to which the site and any connecting roads must be restored following removal of the turbine. Many ordinances require financial assurance in the form of decommissioning bonds, letters of credit, or other guaranties to ensure that developers are held responsible for the ultimate fate of their projects.

- *Permitting processes and requirements.* See below.

Ordinances for large wind energy systems differ further from small wind energy ordinances in that they typically address landscape-scale impacts resulting from turbine construction. Common provisions include the following:

- *Road protection.* Most ordinances require developers to inventory existing road conditions before construction begins and to repair any damage caused during the course of turbine construction.

- *Site clearance.* Some ordinances specify that vegetation clearing and land disturbance during construction are to be kept to the absolute minimums necessary.

- *Soil erosion and sedimentation control.* Soil erosion, sedimentation control, and stormwater management are often addressed by state environmental requirements, though some local ordinances require that appropriate erosion control and stormwater management measures be taken throughout the road and site construction process.

Source: Kern County, California

PERMIT PROCESSING

Though the permitting process for large wind energy systems can be lengthy, it need not be overly complex. Most often, these projects are handled through the conditional or special use permitting process. As with most other large commercial projects, developers are responsible for obtaining the necessary local, state, and federal permits and demonstrating compliance with development standards. The role of the planner is to coordinate this process and make sure that all the application requirements are met, as well as to help both developers and local officials understand the standards.

The permitting process should provide a way to ensure that potential negative impacts are identified, addressed, and mitigated if necessary. Pre-

KERN COUNTY, CALIFORNIA

Kern County, population 839,600, has long been at the forefront of energy production. Throughout the 20th century, the oil and natural gas industries have significantly contributed to the region's economic and industrial growth, and in recent decades Kern County has turned its focus to renewable energy sources, particularly wind. As the home of Tehachapi Pass, one of several identified notable wind resource areas in California, the county is strongly positioned to lead the way in planning for and developing wind energy projects.

Extending from the mountains to the desert in the southern Central Valley of California, Kern County is best known for its concentration of utility-scale wind energy projects; as of 2010, there were 34 wind farms in the Tehachapi wind resource area. According to Lorelei Oviatt, the county's director of planning and community development, large-scale wind energy development first caught the attention of residents and public officials in the 1980s with the introduction of federal and state tax credits designed to spur wind development. The county saw construction of several wind farms at this time, and wind energy development has continued ever since. Oviatt notes that a new era of wind projects materialized in Kern County after 2000, when projects originally built in the 1980s reached the end of their useful lives and more modern and efficient wind developments replaced them.

County planning activities reflect local government and citizen interest in promoting and enabling wind energy development. The 2007 general plan devotes a chapter to energy development activities within the county; a subsection addresses wind energy exclusively, calling for "the safe and orderly development of wind energy as a clean method of generating electricity while providing for the protection of the environment" and outlining specific polices and implementation measures to achieve this goal.

Local land use regulations compliment the goals laid out in the general plan. In 1986, the county added a Wind Energy (WE) Combining District to its zoning ordinance, setting parameters for the development of utility-scale wind projects. The WE district functions as overlay zoning that may be combined with agricultural, industrial, natural resource, forestry, or estate district classifications. In most cases a minimum 20-acre parcel size applies. Detailed development standards for the district address lot size, setbacks, height limits, parking, signs, distance between structures, maintenance and abandonment, and permitting requirements. The zoning ordinance also includes provisions for residential-scale wind turbines; a small wind energy permit system is required for installation, and basic height, setback, and noise restrictions apply.

Local regulations also address the complexity of siting wind energy projects in proximity to military installations. Edwards Air Force Base and the China Lake Naval Air Weapons Station are located in the eastern corners of the county, so a key issue in planning for wind energy here is making sure proposed wind projects will not interfere with these installations. The County has partnered with military personnel to identify locations where wind turbines could compromise military activities; the zoning ordinance includes a military review requirements map delineating portions of the county where proposed structures over specified height thresholds must undergo review by military personnel. Kern County is also within the state's R-2508 Joint Land Use Study (JLUS) area, a cooperative planning effort among military and surrounding communities to achieve compatibility between military mission

Source: Kern County, California

activities and neighboring civilian communities.

To encourage wind energy development, the county uses a permitting process that is time- and cost-efficient for both developers and local government. Once a proposed wind energy project site is rezoned to the WE district—a step requiring both environmental review and public hearings, which can take up to a year—wind energy systems are a by-right use and developers can secure the needed project approvals quickly, with the relevant permits issued over the counter. In addition, a team of county staff members are assigned to wind development applicants, helping to process permits, providing early feedback on proposals, and timing the permitting process to meet other deadlines a developer may have with lenders or investors. According to Oviatt, this team approach is very efficient; it ensures that several staff members are knowledgeable about each project and allows staff to process multiple proposals concurrently. The combination of streamlined permitting and the county's direct approach to working with developers has resulted in a process that minimizes delays and offers a welcome degree of certainty to developers.

Wind energy projects within Kern County have generally met with strong public support, in part due to the host of benefits that wind energy development has brought to many stakeholders. Wind farm construction and manufacturing have boosted business for many local industries and small business owners, private property owners have benefitted from land lease agreements with developers, and local colleges have developed new training programs for wind turbine technicians. Some individuals have voiced opposition to some of the wind energy projects, but Oviatt says that most concerns relate to specific issues that the planning department is able to address with mediation or compromise.

Given the number of wind farms already operating within the county boundaries, it is clear that Kern County is well on its way to realizing the wind energy goals laid out in the 2007 general plan. The growth of the wind energy industry in Kern is not without complications, as increased power generation has raised issues related to interconnection and transmission that must be addressed. Overall, however, the outlook for continued wind energy development in Kern County is positive.

For additional information:

- Kern County General Plan (2007). Part 5.4.2, Wind Energy Development Available at www.co.kern.ca.us/planning/gpe.asp.

- Kern County Zoning Ordinance (2009). Figure 19.08.160, Military Review Requirements Map; Chapter 19.08, Section 19.08.415, Small Wind Energy System; Chapter 19.64, Wind Energy (WE) Combining District. Available at www.co.kern.ca.us/planning/pdfs/KCZOMar09.pdf.

application conferences between developers and key local staff members are recommended. At these, planners can ensure that everyone involved understands the permit requirements and development standards. Developers may not be familiar with requirements particular to certain states or local jurisdictions. Developers and local officials must agree on clear standards for development and expectations for impact mitigation during and after construction, whether through a list of conditions or a development agreement.

Large wind energy systems require significant supporting documentation from local, state, and federal sources. Permitting documentation requirements for these projects may include:

- A site map and plan of all turbine locations, including the locations of structures, roads, utility infrastructure, tree cover, and other significant environmental features

- Landowner lease-agreement documentation

- Environmental permits, which can include erosion control and stormwater management permits from state departments of environmental protection

- Utility interconnection agreements

- FAA approvals and lighting plans

- Highway access permits from state departments of transportation

- Bird, bat, or other wildlife impact studies and monitoring agreements, usually coordinated with the U.S. Fish and Wildlife Service (USFWS)

- Road condition inventory and repair agreements

- Shadow flicker analyses

- Noise studies

- Visualizations or viewshed impact studies

- Decommissioning plans

Perhaps most important, planners should ensure they have the information and resources they need for an informed review of development applications for large wind energy systems. Many of the planners interviewed for the case studies in this report recommended contacting colleagues in other jurisdictions that have successfully addressed this issue to learn from their experiences.

ELEMENTS OF A UTILITY-SCALE WIND PROJECT PERMITTING PROCESS

Environmental Issues

The major environmental impacts to be addressed for a utility-scale wind project include:

- wildlife, particularly birds and bats but also other sensitive wildlife species

- sensitive plants

- habitat, particularly wetlands and other sensitive areas

- changes in water quality or flow that might cause soil erosion or require management (e.g., stormwater controls)

Both the construction and operational phases need to be included in the analysis. The standards to be followed will depend on project size and location. For instance, a single utility-scale turbine requires consideration of its immediate impacts but is unlikely to need a broader study of bird migration. Larger projects need more rigorous analysis, and projects near sensitive species or habitats also need additional study. If federal lands are affected, a more exacting analysis under the National Environmental Policy Act (NEPA) is needed. Otherwise, the federal role is largely advisory as to whether federally listed species might be affected by the project. States may have a process requiring environmental analysis and a determination of no significant impact before approval, but some states have relatively few process requirements. Localities, in turn, vary dramatically in their efforts to protect environmental resources. Some rely heavily on states to deal with the major issues, while others have detailed standards that in some cases duplicate federal or state law.

A potential missing piece is consideration of any local environmental resources that are not addressed by state or federal agencies. If water quality is a particular issue locally, for instance, it is reasonable for the planner to ensure that appropriate information on those impacts be gathered and submitted in parallel with any work done at the state level. This is most likely to be a concern in states where there is no overarching or environmental permit required to build and operate a wind farm.

Still, many state and local agencies are left without a clear framework to approach an environmental evaluation of a proposed wind project. A major piece of guidance for wind projects comes from the USFWS's Draft Land-Based Wind Energy Guidelines (www.fws.gov/windenergy). This is the latest in a series of draft voluntary guidelines that the USFWS has written in conjunction with the industry and wildlife advocates. The USFWS plans to issue final guidance in the near future. The prior guidance has been used as a baseline for environmental analysis of wind projects by a wide variety of actors, from developers and their consultants to local, state, and federal officials charged with evaluating environmental issues with wind projects, as well as by nongovernmental organizations concerned with wildlife issues. Wisconsin's guidelines, for instance, are designed to supplement the USFWS guidelines while drawing attention to issues and resources that are important in the state.

The guidelines take a tiered approach to analysis with an initial screening that puts sites into categories of low, moderate, and higher risk. Each tier then requires a different level of analytical rigor. There are specific recommendations on methodologies to use, which generally have been adopted by the industry as the basic framework of project environmental analysis. There is still dispute about the reasonableness of some specific recommendations. Still, the bulk of the recommendations are being adopted broadly despite their voluntary nature.

All wind projects will go through an environmental screening process by the developer to determine if there are any "red flags" that suggest a site is not developable. Some developers are more thorough than others, however, so it may be difficult for a planner to determine how much work has actually been done on this front in the absence of explicit requirements or disclosure. Virtually all investors and lenders in wind projects today, particularly the large corporations that provide tax equity for larger projects (over $100 million), will require a substantial and rigorous environmental analysis. They want to ensure that the project is not in violation of any law or regulation. They also want to ensure that reasonable efforts have been made to meet commonly accepted standards, in order to avoid or mitigate the possibility of any enforcement action should there be a violation. The USFWS draft guidelines have been widely adopted because they represent a clear standard of care that investors and lenders believe will largely protect them from significant enforcement actions (absent evidence of negligent, careless, or knowingly illegal behavior).

CASCADE COUNTY, MONTANA

Cascade County, population 81,327, is located in central Montana, at the upper reaches of the Missouri River. Well known for its winds—one-third of the county's 2,700 square miles experience strong and predictable class 4 winds—Cascade County is home to more than 400 wind turbines, with more on the way.

Wind energy has had a strong champion in former county commissioner Peggy Beltrone. Though wind was not part of her platform when she became the first woman to serve on the commission in the mid-1990s, the need to diversify the county's revenue streams was brought to the fore when a local energy company protested its tax bill by withholding taxes—$14 million over seven years—on its five hydroelectric dams in Great Falls, the county seat. In 2001, Beltrone visited a wind farm under construction in Pincher Creek, Alberta, and realized that wind energy—and tax income from large turbines sited within the county—had the potential to boost the local economy and help counter the crippling effects of poverty on her region.

Beltrone used her political pull to get others on board. She asked Cascade County GIS technician Eric Spangenberg to develop an electronic map of the county using wind energy data recently developed by the National Renewable Energy Laboratory (NREL). Released in 2002, this wind resources map was a first-of-its-kind effort to accurately overlay the distribution of wind resources on land ownership and other records at a fine scale. Further refinement was provided by other county employees, such as Doug Johnson, then-director of the Weed and Mosquito Department, who contributed his experience with wind conditions and local landowners, gained as his crews sprayed pesticides and herbicides on property throughout the county.

The online wind resources map promised to shave weeks off development time for potential wind energy developers, but Beltrone wanted to actively promote Cascade County as a ripe opportunity for wind energy. She launched a marketing program in 2002, printing and distributing brochures about Cascade County's wind energy development potential using existing staff resources and less than $400 of additional funds. One result of this marketing strategy was an American Wind Energy Association–funded county wind-energy information radio station, advertised by signs on a 50-kW wind turbine powering one of the county's public works facilities.

In 2005, the county—one of the few in the state with a zoning code—developed and approved a comprehensive wind energy ordinance that streamlined the permitting process for developers. The ordinance differentiates between commercial wind energy systems designed to generate power for sale and off-site use and noncommercial systems primarily for on-site energy generation and use. Residential and small wind energy systems of fewer than 50 kW are permitted by right as principle uses in agricultural districts and accessory uses in rural residential, business, mixed use, and industrial districts subject to certain conditions, including setbacks from property boundaries equal to tower height plus blade height plus 20 feet, distancing requirements of 1,000 feet from other

Madelyn Krezowski

structures, noise limitations at property lines ranging from 50 dBA to 75 dBA, and other general safety conditions. Subject to the same conditions (except the property line setbacks), commercial wind facilities of 1 MW or less are permitted principal uses in the agricultural district with special permits required for those more than 1 MW. The standard special use permit requirements—including a preapplication meeting, specific documentation materials, and a public hearing—apply; the ordinance does not specify any special application materials.

Cascade County's efforts yielded results. In 2003, Montana-based Exergy Development Group partnered with a local construction company to build Montana's first commercial wind project at the company's nearby asphalt plant. State assistance helped fund installation of an anemometer tower at the site for data collection, reducing project costs for the developer. Three years later, the six-turbine, 9 MW Horseshoe Bend Wind Park came online, replacing a 3 MW diesel generator that had been powering the asphalt plant. The National Association of Counties (NACo) recognized Cascade County in 2006 with a Sustainable Communities Award for the wind energy marketing program.

Subsequent successes include the selection of Great Falls by Gaelectric, an Irish wind development company, for its North American office, and announcement of a new 100-mile transmission line dubbed the Green Line to help relieve the current electricity bottleneck south of Great Falls. Invenergy, developer of the Judith Gap Wind Farm in Wheatland County (formerly the largest in Montana) recently received county approval of the 16-turbine, 24 MW Big Otter Wind Farm near Belt, Montana; this project has the potential to grow to between 200 and 300 MW in future phases.

(continued on page 96)

(continued from page 95)

Although the momentum for wind energy development in Cascade County continues, signs of waning interest and increasing resistance from certain groups within the county have surfaced. According to county officials, questions about small wind development have tapered off since 2009 due to economic conditions, although interest in large wind development remains. Beltrone left her position as county commissioner in June 2010 to work for a private wind energy developer. Furthermore, recent wind energy development proposals have been met with greater resistance from property owners who fear the loss of million-dollar views. Within the past few months, Cascade County officials have contemplated charging impact fees for wind energy development.

Another challenge to wind energy development has come from the U.S. Department of Defense (DOD). Malmstrom Air Force Base, Cascade County, and the City of Great Falls are examining growing concerns about potential wind energy development near the base in a Joint Land Use Study. DOD wants to produce a "Red-Yellow-Green" map for wind energy development in the county to show areas of military concern. Red represents areas where wind energy development should be prohibited, yellow indicates a need for careful review and potential restriction, and green signals areas acceptable for wind energy development with appropriate review. This type of map currently exists for some military installations in California. Additionally, DOD wants to triple the width of the current launch facility buffer around the base from 1,200 feet to 3,600 feet, within which wind energy development and other incompatible land uses would be prohibited. According to Rick Solander of DOD, public workshops held in February 2011 to educate Cascade County officials and stakeholders about mission requirements and procedures were very productive. The military hopes to involve wind energy developers in future discussions.

Over the last decade, Cascade County has proven itself to be fertile ground for wind energy development, which has brought economic benefits to the area. The Horseshoe Bend wind farm generates $150,000 in property taxes per year, and every 100 MW of installed capacity is estimated to create eight new well-paying jobs. The early flush of development has since been tempered, however, by recent calls for impact fees on developers and pushback by the military and property owners seeking to protect their views. The coming years will tell whether Cascade County's wind industry will continue to grow or if increasing opposition will slow future expansion of wind energy development in one of the nation's windiest places.

For additional information:

- Cascade County Zoning Regulations (2009). Sections 7.2.1.15, 7.2.1.16, 7.2.3.13. Available at www.cascadecountymt.gov/doc/countyzoningregulations2009.pdf.

- Cascade County Wind Power Map. Available at http://gis.cascadecountymt.gov/Website/WindPower.pdf.

- "Cascade County Wind." Available at www.cascadecountywind.com.

- "Cascade County Wind Power: Put Wind to Work for You." Promotional brochure available at www.windpoweringamerica.gov/pdfs/mt_county_commissioners_2007.pdf.

- Grubb, Alex. 2010. "Gaelectric Hopes 'Green Line' Will Help Move Energy." *KRTV News*, December 3. Available at www.krtv.com/news/gaelectric-hopes-green-line-will-help-move-energy.

- Malmstrom Air Force Base JLUS. Available at www.malmstromjlus.com.

- Puckett, Karl. 2010. "Cascade County Zoning Board Oks Belt-Area Wind Farm." *Great Falls Tribune*, December 18. Available at www.wind-watch.org/news/2010/12/19/cascade-county-zoning-board-oks-belt-area-wind-farm.

- U.S. Department of Energy, Energy Efficiency and Renewable Energy, Wind Powering America. 2009. "National Association of Counties (NACo) honors Cascade County as a National Leader in Creating Sustainable Communities." Available at www.windpoweringamerica.gov/filter_detail.asp?itemid=1205. ◀

Cultural Issues

The main concerns in this category are impacts on historic properties and on archaeological artifacts that may be disturbed by construction. Many states have an agency that deals with historic properties, and some have standards to which all new activities must conform. Because of wind turbine height, there will often be some historic properties from which one or more turbines will be visible, similar to cell towers and other large commercial structures. Since moving turbine locations to be completely out of view can be difficult, mitigation may be limited to putting up screens (e.g., trees, fences) to shield a given property from the view.

Aesthetic (Viewshed) Issues

Unless a jurisdiction has explicit laws or regulations controlling its viewshed, this is a particularly thorny issue for planners. Without explicit regulations, there are few objective standards to refer to, and in many ways this comes down to personal preference and property rights. Some believe any changes to the landscape are an affront to their property rights, even if the structure is on not on their property. Others believe that they have the right to do anything on their property as long as there is no physical impact to the neighbor. Sometimes there is a clear community preference, but more often there is a strong minority view on one side or the other.

Most guidance recommends taking aesthetics into consideration in the design of the project, particularly the location and look of ancillary facilities such as storage buildings, offices, transmission or distribution wires, and substations. AWEA's guidelines suggest using turbines and towers with uniform appearance, including color; limiting use of prominent commercial markings; putting power cables underground where feasible; and synchronizing turbine lighting where possible. All of these help to mitigate the visual intrusion. The rule of thumb among developers is that intrusion on the landscape can be accepted by most communities if the benefits outweigh the costs, preferably by a substantial margin, but consensus tends to be difficult to achieve.

Land-Use Issues

Most states or localities have processes for managing the compatibility of new commercial activities with existing land uses. However, feedback from the community can be very useful for planners and developers in figuring out how to create the best balance of interests. For instance, one community that would host a new project had a long tradition of people riding four-wheeled recreational vehicles in farm fields and on some of the hills near where turbines would be. These users were concerned that the guy wires used on meteorological towers would create a hazard. It turned out to be relatively easy to first mark the wires near ground level with colorful markers and then to switch over time to unguyed towers.

Constraint maps are useful tools for mapping out where land-use conflicts might exist and where other concerns (such as environmentally sensitive areas) might exist. Typically, developers map out all constraints in a single map or on layers that can be added or subtracted from a map, including all setbacks (from buildings, property boundaries, transmission lines, roads, microwave beam paths, etc.). Modern GIS mapping capabilities make this a powerful tool for project design and for conforming to and documenting regulatory requirements.

Sound Issues

While sound is objectively measurable, the impact of sound on humans is far more subjective, which gives rise to some difficult issues in the context of a wind project. There are also many different ways to measure sound, and there is no agreement on an objective standard for objectionable sound. Moreover, some turbines are noisier than others, and topography, vegetation, and atmospheric conditions can also affect how sound travels. Sound standards exist in many jurisdictions; where they do not, the U.S. Environmental Protection Agency standard serves as a default.

Jurisdictions with a specific sound standard at a particular location (property boundary or home) tend to rely on turbine setbacks from homes or property boundaries to control noise issues. The typical setback has been 1,000 feet from homes unless the home owner agrees otherwise, but some jurisdictions have adopted greater setbacks, often because of concerns raised by a community. At a certain point, particularly in areas with small parcels, a larger sound setback can essentially block wind projects since the allowable footprint for turbine placement is too small.

Sound issues are difficult to mitigate once a project is built. Adding sound insulation to the nacelle of a turbine can lower the mechanical noise it emits. However, the aerodynamic noise from the blades is often the chief source of complaint, and that cannot be mitigated at the turbine without constraining its operation, such as limiting use to nighttime hours. This can be costly for the operator, who may favor other options, such as compensation. There are several mitigation tools that can be used at residential structures, such as using vegetation screens or using sound-deadening materials, but such measures do not deal with the entire sound spectrum, so their efficacy depends on the sound emitted by a particular turbine and the circumstances of and around a particular structure.

Shadow Flicker

Shadow flicker is created by the movement of a wind turbine blade between the sun and a receptor. With reasonable setbacks, shadow flicker occurs only when the sun is low in the sky and for relatively few hours a year—typically a tiny fraction of daylight hours. Still, for those who are affected, it can be disorienting and annoying. Much like sound, the impact of shadow flicker is highly subjective. Some find it highly troubling while others are not bothered at all, and the recipient's disposition toward the project will largely determine the degree of annoyance. Some claims have been made that shadow flicker can create epileptic seizures, but these have not proved to have any evidentiary basis.[1]

Mitigation options for shadow flicker are also similar to those for sound in that there is little that can be done after construction other than curtail operation or use vegetation or other screens inside or outside an affected house, and compensation to the affected party as appropriate. Some jurisdictions have tried to set standards for a maximum number of hours per year that shadow flicker can occur at a given location, but there is no consensus on what that level should be. Setbacks continue to be the default solution for dealing with shadow flicker, although again there is a great deal of variation on what that distance is. Shadow flicker is reasonably easy to model, with far less uncertainty than with sound, so a planner could ask for a color-coded map showing how many hours the area surrounding a wind project would be subjected to with a given setback. Standards should differentiate between residences and rarely occupied structures or locations such as sheds, fields, or roads.

Safety Concerns

Safety concerns with wind projects usually revolve around the potential collapse of a turbine tower, the "throwing" of a blade, or the slinging of ice from turbine blades. There is now enough operating history in enough locations around the world that reasonable probabilities can be established for these events. The probability of injury or property damage from any of these events is extremely small, but the residual risk can be minimized with reasonable setbacks. The 1,000-foot setback became a standard in part because it creates a substantial margin of safety.

Some jurisdictions have attempted to mandate design standards or certifications, but they must be able to adapt such requirements to a rapidly changing market. As noted, setbacks are often the default protection on safety issues. Those concerned about this issue should focus on the track record and financial solidity of the developer and the eventual project owner, since the owner assumes the liability for operation. In circumstances where the project owner is not known until development is mostly complete, development approval can be made contingent on the owner meeting a certain standard. In any event, such standards are notoriously hard to write in a way that does not create problematic inflexibility.

Virtually all guidelines agree on certain safety recommendations: posting of emergency contact information at various locations throughout the project; posting warning signs about falling ice during winter; locking turbine doors to prevent unauthorized access; and avoiding structures near turbines that would allow unauthorized people to climb them.

Construction Impacts

Construction is sometimes an afterthought in the siting process, but it can significantly affect the immediate community. Construction noise can be managed by setting reasonable hours for the loudest activities. Dust is a frequent source of complaints from any construction activity, so dust sup-

Source: Kern County, California

pression procedures should be discussed early in the process. Developers can be asked for a construction timetable with specific activities, such as delivery of turbine equipment or movement of large cranes, called out to allow various departments of local and state agencies to coordinate their activities. Since schedules inevitably change, good communication among the parties is key to a successful relationship between the developer or owner and the responsible agencies. Good practice also involves notifying neighbors of upcoming road blockages or construction work so that they can work around the constraints. Once again, good communication in advance of and during construction is crucial.

Emergency Conditions

Most local emergency response crews in rural areas have no experience dealing with emergencies in tall structures, much less in wind turbines, so it is important to work out emergency plans covering all reasonable scenarios well in advance of construction. Fires are rare in wind turbines, but they have occurred, and local agencies responsible for emergency response will need a plan for responding to one. Injuries can also occur in the tower or nacelle, so crews require knowledge of how to rapidly get to and move affected individuals in a safe manner. A written plan approved by the appropriate agencies will go a long way toward safe project operation.

Decommissioning Concerns

When a project is at the end of its operational life, turbines and associated infrastructure should be decommissioned and removed. Landowners, communities, and responsible officials will want to ensure that the cost of decommissioning does not fall on landowners or the community. Most projects require the project owner to take responsibility for decommissioning, and some require a financial instrument to ensure completion of the process. Most decommissioning language in easements or in regulations is very general, since as a rule decommissioning bonds have not existed or have been extremely expensive. Thus, "a bond or similar financial instrument" is typically the language of choice. Some jurisdictions have asked for a fund—often a sinking fund paid into over time by the project owner—to be established and held in such a way that it becomes payable to the landowner or the appropriate government agency should the project owner fail to decommission the project in a timely manner. The size of the fund is usually based on the net cost of decommissioning, accounting for a very conservative salvage value of the equipment. Developers and salvage experts maintain that decommissioning can actually make money, since the salvage value of the turbines and related equipment typically exceeds the cost of removal. Still, most jurisdictions want to have some cash on hand to cover other scenarios.

The end of a turbine's operational life is usually defined as occurring once it has "ceased operation" for some defined period, often a year. Since some components have lead times of six months or more, shorter time periods would be problematic for a project owner unless there is an automatic waiver process that can be triggered, perhaps upon evidence of an order.

Some guidelines deal specifically with what standards the removal should meet—for instance, whether all roads should be removed and reseeded. Some specify native plants for reseeding, which is rapidly becoming standard operating procedure for developers. It also can make sense to allow the landowner to elect to keep some or all of the infrastructure items such as roads—or even turbines—since they may be valuable assets.

Economic Issues

Economic issues do not specifically affect the siting and design of the project, but they do inform conversation about the costs and benefits of a project. All economic costs and benefits should be communicated to the public. Some benefits are obvious (taxes, payments to landowners, charitable contributions by the developer to the community), and there are relatively clear construction-period benefits (labor used, supplies purchased, services contracts). Ongoing benefits can include the purchase of materials to repair roads, accounting services, the money spent by maintenance personnel and project contractors in the community on food and lodging, and so on. Costs also need to be addressed, such as who will pay for wear and tear on roads during construction and any repairs involving large cranes and trucks. Communities that rely on tourism are often unsure of the overall impact of a project, so evaluation of similar communities can be very helpful, and some limited studies do exist. While not often covered in siting guidelines, economic issues are still an important backdrop to the community conversation that planners support, participate in, and sometimes orchestrate.

LAND AGREEMENTS

Most developers use easements on property rather than buying land or leasing it. Every developer uses a different form of land easement, which creates some confusion among landowners and any public agencies or other parties that might be interested in representing landowners' interests. Most easements share certain common features, but there is huge variety in what is offered to landowners in financial terms, lengths of contracts, and protections. Landowners often are not aware of standard wind-industry contract terms unless they have done substantial research or know someone who has negotiated a contract, but it is easy to get access to standard contracts via the Internet. Some siting guidelines also available through the web give very specific guidance on what should and should not be in an easement. The Windustry siting guidelines provide a template easement for all parties (Windustry n.d.a).

Some guidelines give specific suggestions for the maximum periods over which evaluation of the project can take place, as well as maximum timeframes for the operational period. Others simply point out the pros and cons of different approaches. The most landowner-friendly guidelines suggest that all information in easements should be made public, which

Source: Kern County, California

has real benefits but can create a difficult dynamic. If a developer is trying to accommodate particular concerns of some landowners, every concession that a developer makes then becomes a matter of public record, which could make the developer less likely to agree to concessions. If some landowners have more valuable assets than others and thus are paid more, it can create tension among neighbors. Jurisdictions that are considering a requirement to make easements public should be cognizant of the pluses and minuses.

Providing a template agreement to landowners can have educational value, but requiring the use of a preapproved agreement can be problematic. Every developer has its own view of what kind of easement can be financed. In fact, some developers get preapproval of their easements from investors or lenders to ensure there are no surprises. The language in an easement, however, changes over time based on judicial decisions and on the results of recent financings. Providing a required easement will almost certainly require the developer to come back and renegotiate the easement, something that neither the developer nor most landowners want to do.

There are some central points regarding easements that many guidelines agree on:

1. Evaluation, feasibility, or option periods should be limited to a reasonable time. Five to seven years is common and sufficient for most of the United States, but it is not uncommon in some parts (the Northeast and California, for example) for it to take that long or longer to get a project ready for construction. An additional option period is not unreasonable but should also have a higher compensation level.

2. Landowners should be aware that standard operating periods are 30 to 40 years in most locations—longer periods do not need to be prohibited but should receive additional scrutiny.

3. Landowners should not have legal liability for anything done by the developer or project owner and vice versa, so explicit indemnities and appropriate insurance requirements should be included.

4. If production-based payments (royalties) are part of the easement, the landowner should have the right to see documentation of the production and have a representation from the project owner that the records are accurate.

5. Decommissioning should be the responsibility of the project owner (unless a landowner explicitly agrees otherwise), and language to that effect should be in the agreement. A specific pledge of funds for decommissioning from the project owner is appropriate but should also be reasonable. If the owner should fail to remove the facility, any decommissioning fund should be payable to the landowner first and then to the jurisdiction, should the landowner not take action.

At least one guidebook suggests that wind rights (and any income associated with them) should not be severable from the land. It is legally complicated to separate wind rights and to ensure that that fact is properly communicated to the next owner. It can also create unforeseen problems for future owners.

BONDS AND GUARANTEES

While bonds and other financial guarantee mechanisms are not commonly used during the development process, they can become relevant during the construction and operational phases. Language requiring such financial mechanisms should be broad enough to allow flexibility in how the funds are provided, since bonds may be expensive or not available.

In the construction or preconstruction phases, localities typically will require a bond to be posted to cover potential road damage from the heavy traffic associated with a wind project. The roads department or its equivalent usually works with the developer's construction plan—and any related permit submissions—to determine the timing and likely impact of construction-related activities on the roads. Once an amount for road repair has been determined, the developer usually provides a bond or cash payment. The locality will then hold that amount until shortly after construction has finished and road repairs have been completed. Additional amounts might be required if construction is to take place during periods of frost or high precipitation, or if truck weights are near the design limits of the roads. More frequently, movement of heavy loads will not be able to begin until after the frost period.

Broader construction bonds could be utilized for special circumstances such as protecting sensitive resources (e.g., drinking water), if those are potentially threatened by construction activities. In this case, bonds might be structured to cover remediation costs.

During the operational phase of the project, bonds or other guarantees are not common except for any support of decommissioning requirements. As discussed, decommissioning bonds are often not commercially available, so other financial mechanisms (sinking funds, cash deposits, pledge of other assets) should be allowed.

THE PROCESS OF DEVELOPING A WIND ENERGY PROJECT

Larry Flowers and Dale Osborn

With the growth of the wind industry, planning and zoning officials, legislators, and other elected officials are increasingly confronted by advocates for and opponents of the siting of wind turbines. This section defines the issues that planners may confront when a wind project is proposed, and it explains planners' relationship to the preconstruction development process of wind developers. With this information, planners can better understand how and

COMMUNITY WIND PROJECTS

Larry Flowers

Community wind projects typically range from a single turbine up to several tens of MW. There are a wide variety of community wind applications, including electricity-use reduction on a farm or ranch or in a business, school, or community facility; diversification of supply for a rural electric cooperative or municipal utility; and sale by a community-based independent power producer (IPP) or limited liability corporation (LLC) to a local electricity supplier. AWEA estimates that 5.6 percent of the utility-scale wind market at the end of 2010 consisted of community wind projects.

A number of innovative business models and policies have emerged over the last decade, designed primarily to address the inability of many individual investors (i.e., locals) to effectively utilize the Production Tax Credit (PTC), which requires the use of passive income. (See page 15.) The Minwind model (see page 21) arose from a group of Minnesota farmers who formed an LLC based on the ethanol co-op model. The "Minnesota Flip" was designed to make use of the PTC appetite of larger companies and then flip the majority ownership to locals after 10 to 12 years, once the PTC was fully monetized. Several variations on the Flip model, in which an organization with sufficient tax appetite joins ownership with local investors, have emerged. In Colorado, the "piggyback model" was employed to take advantage of the economics of scale in manufacturing, construction, and operations of a nearby conventional large project. (See page 60.) South Dakota has taken the piggyback structure a step farther and has sold shares to state residents of the minor portion of a large project. Some innovative lease structures allowed under the Investment Tax Credit (ITC) / cash grant federal incentives have also emerged, as wind projects have been eligible for these incentives since 2009. Any project owned by a rural electric cooperative (REC) or municipal utility qualifies as a community wind project, since the utility owners are local; these projects are often financed by conventional public power sources.

The main benefits of community wind projects that are driving the market are increased local economic development and local control. When at least some of the owners are local, a greater percentage of the project revenues flows to local people, businesses, and institutions. The direct, indirect, and induced income streams from construction and operations can be up to three times those of conventional out-of-state-owned projects (Lantz and Tegen 2009). The actual magnitude of increased economic benefits depends on the availability of local, qualified labor and material supplies, as well as on the ownership structure and financing details. Local control also often helps in siting and sizing projects to suit local interests.

Other benefits include the sense of community created when a place produces some of its own energy, as well as the feeling of independence that comes with it. People may also feel that they are contributing to improved local (and global) air quality, as well as protecting the local watershed through water savings.

Because of community wind projects' smaller scale, they sometimes do not require expensive and time-consuming transmission upgrades, and often they can be connected to the distribution grid easily. The smaller project size can reduce the possible environ-

Andrew Stern

mental impacts on wildlife and scenic views; however, community wind projects still must abide by local ordinances, as well as state and federal environmental regulations.

As in conventional ownership structures, both federal and state policies are important drivers. While the PTC has had limited value to community wind projects, the recent U.S. Treasury grant was particularly effective in helping community wind projects remain competitive. The 30 percent ITC has broader application to community wind projects, as locals can make better use of it than of the PTC. In addition, the U.S. Department of Agriculture's Rural Energy for America Program (REAP) has been effective in reducing the predevelopment and installation costs of rural community wind projects through grants and guaranteed loans.

At the state level, there are 29 states with standards requiring utilities to include a certain amount of renewables in their portfolios, but these do not specify a particular level of commitment to community wind projects. Minnesota led the way in community wind projects by requiring Northern States Power (now Xcel Energy) to purchase 100 MW from community wind projects of 2 MW or less at a reasonable tariff, combined with production incentive payments. Several states have incentivized community wind projects through community-based energy development (CBED) legislation.

While siting of community wind projects should be less challenging than that of larger, conventional projects, community wind projects do have some distinctive features of which the planner should be aware:

- They are partially or wholly owned and controlled by local companies, organizations, or citizens.

- They provide for up to three times more local economic development benefit than conventional, third party–owned projects.

- They are usually smaller than conventional projects.

- They often partially serve the electricity needs of the community.

- They are often connected to the electrical distribution line and do not require a separate overhead transmission line.

Projects still need to meet local codes and ordinances, satisfy federal and state rules and regulations, and incorporate good siting practices—particularly early, frequent, and responsive community involvement. The standard issues of economics (especially rate impacts), aesthetics, setbacks, sound, wildlife impacts, property values, radar, infrastructure impacts, impacts on cultural and historical resources, and land use apply to community wind projects as much as to any other. A particularly thorny issue is the impact assessment of community wind projects on avian and bat species, for which there is little data. (See also page 37.)

Community wind projects are often sited close to communities and thus are often in view of owners and nonowners alike. It is reasonable to expect less resistance to a community wind project

(continued on page 104)

(continued from page 103)

from those connected to it. However, there are always individuals or groups who resist change, especially to the landscape. Diversity of opinion within a community can be a delicate issue for local planners and officials.

Three challenges to community wind projects are competitive economics, utility resistance to purchasing their electricity, and siting requirements and regulations. The generally smaller size of community wind projects makes it harder for them to compete with conventional 50–400 MW projects in a competitive market. Additionally, utilities seem more inclined to make one or two large purchases to meet their needs or requirements, instead of many smaller purchases. And there is a movement among investor-owned utilities to own their own wind projects, rather than purchasing power from a third party.

Public power producers (co-ops and rural municipalities) seems to be the perfect fit for community wind projects, as they are also locally owned, benefit from the economic development of community wind projects, and often have smaller loads and growth. But most of them are not subject to their state renewable-energy portfolio standards requirements and often see wind as too expensive.

Interest in community wind is spreading across the country as wind stakeholders seek more direct involvement in the wind energy future. Local business leaders and officials see community wind projects as means to generate economic development while stabilizing energy prices and improving the local and regional environment. Policies on siting, ways of improving competitive economics, and developing utility demand are all necessary for community wind to meet its enormous market potential. ◀

when local planning and development review procedures interact with the wind development process.

There are five key activities associated with the successful development of a wind energy facility: (1) acquiring land rights; (2) completing wind resource studies; (3) obtaining environmental and land-use permits; (4) studying, analyzing, and obtaining transmission rights; and (5) completing a power purchase agreement (PPA) or a facilities sales agreement. Planning and zoning officials may consider each of these when contemplating ordinances and regulations, but some of the specific details are the focus of state and federal agencies; duplication of effort may not be worthwhile.

Acquiring Land Rights

Developers and the finance community require that specific rights be granted by a landowner for the development, construction, operations and maintenance, and reclamation of a wind energy facility. The key ones are the rights of ingress to and egress from the property; rights to study the wind resource and construct and operate the wind facility for a specific term (generally greater than 35 years); and the right to collect and transmit off the property the electricity produced by the facility. For these rights, the landowner is compensated by the developer. Since these agreements are between the developer and the landowners and result in no new built structures or changes in land use, public officials often have very limited influence over them. However, some state governments have enacted legislation to define certain terms of these agreements.

Wind Studies

Authoritative wind studies require the installation of calibrated wind speed and direction instruments. These instruments are placed on a meteorological tower typically 60 meters (196 feet) high and require no excavation or concrete, as they are supported by guy wires and screw-in anchors. Depending on the size of the project area, multiple towers may be warranted.

Some counties have enacted permitting requirements for these towers. The requirements may include aviation marking, obstruction balls, and in some cases lighting, which may require incremental power supplies. Towers over 200 feet tall require a "determination" from the Federal Aviation Administration (FAA) as to what lighting is required.

As long as these towers are located at least 200 feet from a property line, they generally have minimal impact on neighbors. Some counties have implemented an environmental permitting process for these towers. A building permit is also usually required.

Environmental Permits

The scope of environmental permitting is heavily dependent on the ownership of the land being developed: private, state, or federal. If the project is located on private or state lands, the state agency that oversees the protection of sensitive plant and animal species should be consulted by the developer early in the process. Often, these agencies will request or require that certain data be collected before rendering an opinion on the environmental suitability of the site. If federal land or funds are involved, federal regulations (e.g., NEPA) or agencies (e.g., the U.S. Fish and Wildlife Service) will play roles. On the specific cases of avian and bat impacts, see Chapter 3.

Developers will customarily hire qualified biologists and botanists to conduct a preliminary site assessment known as a fatal flaw analysis. These analysts will review the area, identify potential sensitive species within it, and interview local experts. This study provides the developer with the necessary tools to work with both state and federal enforcement agencies to determine what further studies are needed. The developer will conduct these studies and report periodically to the enforcement agencies. Similar studies may

TOWN OF HULL, MASSACHUSETTS

Glen Cooper

Hull, Massachusetts, a town of 11,000 residents, is home to two large municipally owned wind turbines. Hull I is a 660 kW turbine commissioned in 2001; Hull II, a 1.8 MW turbine, was commissioned in 2006 and sits on top of a former 13-acre landfill, the first such siting in the nation. The turbines are owned and operated by the local power utility, Hull Municipal Light Plant (HMLP), and produce approximately 10 percent of Hull's electric needs. An urban and coastal environment—Boston is only 10 miles away across the peninsula—makes these wind projects distinctive.

Hull has a long history of harnessing wind energy; community members have referred to the Hull I site as "Windmill Point" for nearly 200 years. In the 19th and 20th centuries, wind was used in this region to power mechanical tasks, such as pumping water and grinding grain. Windmills have occupied what is now the Hull I site since the 1820s, pumping water that was then evaporated to yield salt used to preserve and pack fish. Modern efforts to harvest wind energy in Hull began in the 1980s when a small wind turbine was installed near the local high school to offset the school's electricity use. This turbine functioned effectively for years, reducing the high school's electricity bills by 30 percent, but suffered irreparable damage in 1997 after a heavy storm. Shortly thereafter, local stakeholders began investigating the possibility of repowering the site with a utility-scale turbine owned and operated by HMLP.

The community began the planning process for the Hull I turbine in 1997. From the beginning, interested residents were a significant part of the effort; they formed the group Citizen Advocates for Renewable Energy (CARE) and advocated for HMLP's involvement. In 1998, the University of Massachusetts at Amherst's Renewable Energy Research Laboratory (RERL) and the Massachusetts Division of Energy Resources conducted a detailed technical analysis for a potential new turbine at Windmill Point. Following the positive results of the study and growing media interest, HMLP, CARE, RERL, and other stakeholders presented the proposal at a townwide public meeting in 2000. Apart from one opposed resident, public opinion was strongly positive, so HMLP went ahead. By the end of December 2001, Hull I was built and generating power. A similar cooperative process was followed in developing Hull II, which came online in May 2006.

In Massachusetts, municipal light plants are exempt from zoning requirements, and Hull's zoning bylaws do not address wind energy. Documentation required before turbine construction could begin included FAA permits, New England Power Pool (NEPOOL) Interconnect System Impact Study Agreements, and Massachusetts Environmental Policy Act (MEPA) permits and Environmental Impact Study (EIS) reports.

The Hull community wind energy projects have received strong public support and are considered successful examples of community wind energy installations. In 2007, Hull won the U.S. Department of Energy's Wind Power Pioneer Award, which commended the Town's outstanding leadership in advancing wind power and engaging the entire community in its wind power projects. Additional wind energy development may be in Hull's future; HMLP, RERL, and the Massachusetts Technology Collaborative have proposed a four-turbine, 12–20 MW offshore wind farm that could provide 100 percent of Hull's energy needs, though its future is uncertain due to escalating cost projections.

For more information:

- Hull Municipal Light Plant. "Hull Wind Turbine Information" and "Hull Wind Offshore Presentations." Available at www.town .hull.ma.us/Public_Documents/HullMA_Light/light.

- Hull Wind. Available at www.hullwind.org.

- Manwell, J. F., et al. 2003. "Wind Turbine Siting in an Urban Environment: The Hull, MA 660 kW Turbine." Presentation at the American Wind Energy Association Windpower conference, May. Available at www.ceere.org/rerl/publications/ published/2003/AWEA_Hull_2003.pdf.

- Manwell, J. F., et al. 2006. "Hull Wind II: A Case Study of the Development of a Second Large Wind Turbine Installation in the Town of Hull, MA." Presentation at the American Wind Energy Association Windpower conference, June. Available at www.ceere.org/rerl/publications/published/2006/ AWEA%202006%20Hull%20II.pdf.

**WORKING WITH WIND ENERGY DEVELOPERS: ADVICE FROM A
WIND ENERGY INDUSTRY EXPERT**

Kevin Rackstraw

This section is designed to help planners understand the perspective and motivations of developers in the hope of engendering better communications that will be beneficial to both sides of the relationship.

The prospect of profits can be a source of tension between community members (including planning officials) and developers. Wind development is a costly and multiyear process. It is not unusual for a developer to spend three to five years in development and to accrue between $2 million and $5 million in development costs. It is commonly accepted in the industry that a developer might spend $1 million to $2 million fully developing a project in relatively easy-to-permit areas, whereas in more complicated areas such as California or parts of the East, a developer can spend double or triple that amount. Generally, the greater the population and the larger the number of competing land uses in a community, the longer and more expensive the process will be. However, projects located closer to loads (users) can often be more profitable than ones in remote areas.

It is important to recognize that most development costs are "at risk"—meaning that there is no guarantee of any return of invested capital until the project is sold or achieves long-term financing—until financing is committed. That in turn typically requires all permits in hand, all land control complete, turbines on order (or at least reserved), a construction firm committed, and a contract to sell power from the project over 10 to 20 years. Many projects never see the light of day even after millions of dollars have been spent on them. This is why most development is completed by large and well-capitalized firms that can afford to wait through months or years of inevitable delays. This also helps explain why developers are reluctant to give up on a project. Projects that do make it to the finish line can appear to be quite profitable, but they need to be in order to make up for the many that miss. It is not uncommon for developers to have three to five projects in the works for each one that is finished.

At its best, development is carried out with early and extensive communication among all affected parties. However, there are several factors that tend to inhibit open communication between developers and planners or regulators at the local, state, and federal levels.

- **Competition among Developers**

  Developers create the most value by finding an area with good wind resources and then developing it at the lowest reasonable cost. Competitors for a site bid to sign landowners and create a race to get control of a site. Nobody likes spending tens or hundreds of thousands of dollars in the early stages to find out that someone else has swooped in and captured site control. For this reason, developers tend to like keeping a low profile until they have confidence they have a good site.

- **Sensitive Information**

  Developers are loath to give up certain kinds of information, particularly wind resource data, cost of energy projections, and development costs. Asking for that information is like asking oil prospectors where they have found oil, what the oil's quality is, and how much it will cost to get it out of the ground. Making such information public is tantamount to handing it to competitors.

- **Uncertainty about Project Impacts and Design**

  When a developer first gets interested in a site, usually very little is known about it. Until the developer has more knowledge of the site, in particular any red-flag issues, it is reluctant to be in the public eye. This may manifest as a reticence to come before planners and other government representatives. Most developers like to speak from a position of knowledge in order to engender confidence.

(continued on page 107)

be conducted for issues related to archeological and historic preservation.

In most cases, the local permitting authorities are not directly involved in such studies, but they may require the results from the studies as part of the local permit application. They may occasionally require a letter from the state agency commenting on the project. Local planning and zoning ordinances often contain language requiring developers to address environmental impacts.

Land-Use Permitting

Private land is the purview of local officials. Some jurisdictions have recognized the economic development potential of wind energy. In many rural counties, even a project of fewer than 15 wind turbines can be the largest property tax payer. Many windy cities and counties have enacted siting ordinances. In addition to economic development, local officials are concerned with the health, safety, and concerns of their residents. Consequently, their regulatory focus is on the impact of wind turbines from sound, proximity to neighbors, site access and construction traffic, viewshed, shadow flicker, and property values. (See Chapter 6.)

In consideration of a land-use application, local officials should evaluate the planned interconnection routes to the transmission system. (See Chapter 4.) In large projects, these "generation feeders" may be many miles long and may cross multiple counties. These lines are subject to state or federal environmental review, but local officials will also need to review the application. Setbacks are typically not required for such lines. For lower-voltage "collection" systems, these lines are usually below ground and should be buried deep enough to avoid conflicts with utilities, farm plows, and other land uses.

(continued from page 106)

Another area of uncertainty is project design. Initial locations of turbines, as well as other project infrastructure such as substations or roads, are often just guesses and will be modified numerous times based on wind resource assessment, environmental constraints, setbacks, land control, and so on. Developers tend to keep project layouts close to their vests until they need to disclose them in permit applications. A preliminary project design can create expectations that might not be fulfilled, create unnecessary rifts among neighbors, or create anxiety about a location. This reticence, of course, has to be balanced with the public's right to information, and good developers will work to ensure that affected neighbors are compensated in some way or otherwise have their issues addressed.

- ### Uncertainty's Impact on Developer Cooperation

  There is typically some tension between a developer, who wants to avoid spending money and time, and a planner, who wants to ensure that standards are met or procedures are followed. Having a clear approvals process in place with specific deadlines for response helps mitigate this tremendously. If the approvals process is ill-defined, schedules and budgets can be unpredictable.

Incentives

Once a project is determined to be viable, well-capitalized developers are often willing to spend money to meet requirements for approval. However, developers will always keep an eye out for ways to do things faster as well as cheaper. One of the main reasons that developers are concerned with speed is that the Production Tax Credit (PTC; see Chapter 2) usually runs on a one- to three-year cycle. A developer does not want to be millions of dollars into a project and have the main tax credit expire, with no certainty that it will be extended on the same basis. Expiration of the PTC can subtract millions of dollars of value from a project—often tens of millions for any project above about 50 MW. Thus, at the end of a given tax credit cycle there is a big push among developers to get projects constructed. This creates tension with longer-term processes such as environmental studies that can extend for more than a year. Good developers will not cut corners, but the pressure to get a project finished in a given year when the PTC expires can be significant. This is another reason why developers should be well capitalized so they can wait out these tax-credit extensions when necessary.

The Value of Uniform and Well-Organized Documentation

Good developers create project assets that do not have messy project documentation with lots of special and sometimes unclear conditions. Investors and lenders want projects that have uniform documentation that can be evaluated easily. To manage risks and be able to create a valuable project asset in an easement, for instance, developers need consistency in the agreement language. This creates a disincentive for developers to negotiate every detail in an agreement, as it would make it more difficult for an investor to understand. It is also difficult to complete land negotiations when landowners believe that every item is negotiable.

The same thing applies to permits. Smaller localities sometimes have an informal permitting process. However, banks and investors have a very hard time accepting any permit or other documentation that is not clear, specific, and official. When a developer says that a bare-bones permit is not sufficient, he or she is not being disrespectful of local traditions but merely recognizing that it is less painful and expensive to create proper documentation at the outset.

Source: Kern County, California

(continued on page 108)

(continued from page 107)

Recommendations for Working with Developers

If there are no wind projects in development or proposed in your jurisdiction, take time now to develop your guidelines for them. Work toward a clear approvals process with a timetable for approvals from your agency and any others you coordinate with. This will set a good foundation for interacting with developers. Developers are appreciative of good, solid process even if it includes particulars they do not like.

Coordinate your process with other agencies at the local and higher levels. To the degree possible, try to avoid duplication of efforts by accepting information submitted to another agency. This can help both the developer and your own staff, since there has usually been another set of eyes (or two or three) that has reviewed the material, particularly if it is a larger jurisdiction, such as the state.

If there are wind projects in development in your jurisdiction, it is not too late to develop a good, predictable process. Do not hesitate to ask developers how to accommodate their interest in streamlining while meeting your obligations for thoroughness and integrity. Developers will accept reasonable efforts to establish good process. You may not accept what they want to do, and you should tell them that, but it will give you valuable insight into how they view your existing process, as well as into what has worked in other jurisdictions. Convening a group of developers and other stakeholders (optimally with an experienced mediator) may be the most effective way of getting valuable input for developing or modifying a process. It sometimes takes a while for such a process to get moving as you build trust with the participants. Check with your counterparts in other jurisdictions that do have wind projects to see what has worked and not worked for them.

Establish a process for protection of confidential information. Even if developers are not asked for such information, it helps build trust in you and your agency if a formal process is in place. This is another area where it is useful to ask developers what they consider confidential and would need protected. This avoids potential conflict at a later and perhaps more time-constrained point.

When specific requirements are being planned, particularly design-related items such as setbacks, get feedback from developers, consultants with project experience, and counterparts who have experience with wind development. There are scores of examples of seemingly reasonable and well-meaning requirements that turn out to be unworkable or even challengeable in court. It is much harder—and destructive to a cooperative relationship with developers—to change requirements after they are announced than to get good input in advance.

Try to give developers a heads up on any significant process changes or requirements in the works. That could make your job more difficult since they might try to lobby against the changes, but in most cases it should build trust. Regular, open communication will encourage developers to give you a heads up as well when things change on their side. ◀

Some municipalities require a reclamation bond before issuing a permit, so that if the project stops or never starts operation, the municipality may remove the equipment. However, in the last 15 years no commercial wind facilities have been abandoned in a way that reclamation funds were needed.

Other Reviews and Requirements

Radar. In considering the application for a hazard determination for a wind project, the FAA evaluates proximity to airports, military flight paths, and Department of Homeland Security radar installations. Typically, if the project is closer than six nautical miles from an airport of any size, a negotiation may be required. Smaller community airports tend to be more flexible than commercial ones, while military constraints tend to be specific to the site and mission of the facility.

Interconnection and Transmission Studies. Three studies are required by the Federal Energy Commission (FERC): (1) a feasibility study; (2) a system impact study; and (3) a facilities upgrade study. (See Chapter 4.) FERC requires a deposit for these studies of $165,000 for a project of 20 MW or more.

ENDNOTE

1. According to the American Epilepsy Foundation, the range of frequencies most likely to cause epileptic seizures is 5 to 30 Hz. The foundation recommends that epileptics keep any exposure to flashes to less than 2 Hz (AEF n.d.). Shadow flicker from wind turbines has a frequency between 0.5 Hz and 1.25 Hz (Noble Environmental Power n.d.).

CHAPTER 8

Lessons Learned

Suzanne Rynne, AICP, with Erica Heller, AICP, Ann F. Dillemuth, AICP,
Joe MacDonald, AICP, Kirstin Kuenzi, and Anna Papke

This report includes 20 case studies from across the country that were developed based on extensive research and interviews with individuals involved with those projects. While their stories have been told separately throughout the report, this chapter draws some overarching lessons in planning for wind energy from their collective experiences, looking at common themes, similar lessons learned, and repeated observations.

In general:

- *Learn from other planners who have experience with wind energy projects.* Many planners who now have significant expertise with wind energy projects learned by forging new paths in the early years of wind energy development. Today, planners who are encountering their first wind energy project application can and should learn from these pioneers. Contact planners in jurisdictions that have already addressed this land use to ask for advice; in most cases, they will be glad to share the lessons they have learned.

- *Address wind energy systems in the comprehensive plan.* The first proposal for wind energy projects can be difficult, especially if there are no standards for this use in place. A statement of support for alternative energy technologies, including wind energy, in the comprehensive plan can help bolster justification for approving this unfamiliar use for the first time, as planners for Mahanoy Township, Pennsylvania, found when faced with the first development application for wind energy the area had seen. (See Table 8.1.)

Source: City of Greensburg

- *If your community does not yet have wind energy standards in place, it is better to adopt an ordinance proactively than to be caught unprepared.* There are many model codes and good sample ordinances from communities large and small that planners can use to help draft wind energy standards that fit their communities' needs. It is better to have a thorough discussion of this issue and put standards in place to be ready for any future wind energy development proposal than to be unprepared and have to scramble to figure out how to address this use the first time a wind energy developer proposes a project. Having a wind energy ordinance in place also signals to developers the level of community acceptance when deciding whether to explore potential projects and gives them better guidance in developing a successful wind energy project application.

- *Partnerships between local governments and nongovernment experts can be extremely helpful for the development of wind energy regulation and development planning.* The case studies highlight not just what the local government and constituents were able to accomplish but also the successful

TABLE 8.1. ZONING TREATMENT OF WIND ENERGY DEVELOPMENT IN CASE STUDY COMMUNITIES

Community	Large (Commercial) Wind		Small (Residential) Wind	
	By-Right	Special	By-Right	Special
Anchorage, Alaska[a]		X	X	X
Appalachian State University/ Watauga County, North Carolina		X	X	
Cascade County, Montana[b]	X	X	X	
Gratiot County, Michigan[c]		X		
Greensburg, Kansas[d]			X	
Hays, Kansas[e]				X
Hull, Massachusetts[f]	X		X	
Kern County, California	X		X	
Lamar, Colorado	X		X	
Madison County, New York[g]		X		
Rockingham County, Virginia		X	X	
San Bernardino, California	X		X	
Washington County, Maryland[h]			X	
Washoe County, Nevada[i]	X		X	X

Notes:

[a] The Anchorage Municipal Code permits single small Wind Energy Conversion Systems (WECS) in designated rural and residential districts. However, two or three small WECS may be installed only in designated mixed and industrial districts as a conditional use.

[b] Cascade County permits small WECS in designated districts and single large WECS in agricultural districts. However, large wind farms may be installed in agricultural districts only with a special use permit.

[c] Gratiot County ordinance requirements apply to all Wind Energy Conversion Facilities, which shall be permitted as a special use in a Wind Energy Facility Overlay District. Small wind facilities are not addressed in the county's ordinance.

[d] Greensburg's Sustainable Land Development Code permits small WECS in all zoning districts that permit structures. However, the Greensburg regulations do not address large WECS.

[e] Hays allows WECS as a special use throughout the community, but strong height restrictions (125 feet in nonresidential areas and 45 feet in residential areas) effectively prohibit utility or community wind.

[f] Hull's community wind projects at the municipal power plant are not subject to zoning regulations. However, the State of Massachusetts developed a Model As-of-Right Zoning Ordinance or Bylaw: Allowing Use of Wind Energy Facilities in 2009 (www.mass.gov/Eoeea/docs/doer/gca/gc-model-wind-bylaw-mar-10-2009.pdf).

[g] The Town of Fenner in Madison County granted utility-scale wind turbines special use zoning permits through a series of public hearings.

[h] Washington County's Sustainable Land Development Code permits small WECS in all zoning districts. However, the Washington County ordinance does not address large WECS.

[i] Washoe County's code permits small wind energy systems by right, so long as the systems meet certain output, height, and lot size restrictions. Otherwise, a special use permit is required. Commercial-scale wind is permitted in designated districts.

Richard Vander Veen, Wind Resource, engaging citizens in a wind-related conversation, June 2011.

Kerry Battle

interaction between local officials and other community leaders: academic leaders and students from nearby institutions of higher learning, national wind energy experts, and collaborative systems advocates. Wind energy development may be a big change for communities, and the insight and support of nongovernment institutions and individuals who recognize wind energy's value for the community can help.

- *Allow opportunities to hear and address public concern.* Efforts to get all people to and all issues on the table mean fewer surprises and less resistance down the road. Dialogue among a wide array of stakeholders provides early opportunities for people to raise concerns and for developers and officials to address those concerns. For large wind energy projects, it is important to provide these opportunities on a project-by-project basis; for small wind energy, planners should reach out and seek input during the policy-setting and regulation-development stages.

- *Wind energy is not a new concept.* Wind has historically been captured to power sailboats and machines such as grain grinders and water pumps. Modern science in the 20th century has sparked new uses for wind, and it has emerged as the fastest-growing source of energy in the world. Technological advances have improved turbine capacity and performance, bringing this form of clean energy and its benefits to the forefront of a culture increasingly focused on sustainability. Wind turbines hold out the promise of a reliable, renewable energy source for the foreseeable future.

For utility-scale wind energy development:

- *Early and thorough communication between staff and developers is key.* Due to the scale and complexity of large wind projects, it is important for planning staff to begin conversations with wind project developers as early in the process as possible and to pull together a team of staff from other departments, such as public works, that will be involved in the permitting or inspections process. This benefits all parties: developers will leave with a better understanding of the requirements for project approval, and staff can help shape proposals to better mitigate any potential negative impacts.

- *Interdepartmental coordination is important.* Again, due to the scale and complexity of large wind projects, several local government departments are likely to be involved in the permitting and inspections process. Forming an interdepartmental team of key staff helps keep all members informed of the development proposal's progress and better ensures that nothing is overlooked.

- *Wind energy can help local economies.* Large wind energy projects can provide economic value to communities in a variety of ways. Due to the current nature of electricity and the grid, the energy produced by a utility turbine does not always stay just in that community, but property taxes and lease payments to owners of land where turbines have been sited do. Construction and maintenance of wind farms also creates both short- and long-term jobs, and wind farms can also generate tourism dollars, as in the case of Fenner, New York. And through alternative ownership structures such as limited liability companies (LLCs), local residents can partner to develop and manage utility wind projects, as has been the case with Minnesota's nine farmer-owned Minwind projects.

For small-scale, residential wind energy development:

- *Adopt small-scale wind energy systems standards.* Unlike for large wind where negotiated agreements are the norm, for small wind energy adopting written standards is an effective way to ensure that projects are compatible with their surroundings. Potential nuisance impacts are well understood and can be effectively controlled through appropriate regulations.

Source: Kern County, California

- *Ensure that adopted regulations actually allow functional wind energy systems to be built.* Small wind energy projects can be extremely sensitive to minor changes in location and height. Allowing appropriate variations for height and siting for small wind turbines is essential to ensure adequate access to wind and effective harvesting of energy. Planners should understand that adopting an otherwise solid small wind ordinance that overly constrains

Source: Kern County, California

height is actually a long, time-consuming way to say no to wind energy development.

- *Set up a straightforward, standardized application process.* If a permit is required for small-scale projects, the permitting process does not need to be complicated, as long as small-scale wind systems comply with basic standards for safety. For small wind projects, complicated and time-consuming permitting processes can add up to 10 or 20 percent to total project costs. In Kittitas County, for example, planners set up a user-friendly, over-the-counter permitting process, making it easy for residents to obtain permits through submitting the required information.

Results of the American Planning Association Survey of Current Practice, Challenges, and Resource Needs

Anna Papke, Ann F. Dillemuth, AICP, and Suzanne Rynne, AICP

As part of the first phase of developing this PAS Report, the American Planning Association (APA) conducted an informal survey of its membership to assess the current state of wind energy planning in communities across the country; discover what challenges planners are facing in planning for, regulating, and implementing wind energy facilities; and ask what information or resources would be most helpful to them in planning for wind energy.

SURVEY BACKGROUND

The survey was conducted online using Zoomerang. It was targeted at planners who are grappling with wind energy planning in their work or communities. APA launched the survey on July 2, 2010, and announced it to APA's approximately 40,000 members in the July 6 edition of *APA Interact*, APA's bimonthly e-newsletter. An invitation to take the survey was also posted on the main page of the APA website, as well as the project page, and was e-mailed to 61 planners who had attended the Planning for Wind Energy facilitated discussion session at the 2010 National Planning Conference. The survey remained open throughout the duration of the project.

SUMMARY OF FINDINGS

Attitudes and Experience with Wind

- Over four-fifths of respondents reported positive attitudes toward wind energy, though some expressed reservations about potential negative impacts of this technology and the complexity of the issues involved.

- Three-fifths of respondents estimated that public opinion toward wind energy in their communities was more positive than negative, though not overwhelmingly so.

- Over two-thirds of respondents had worked on a wind energy ordinance; over half reported that their communities had drafted ordinances, with just under one-quarter of respondents reporting visioning and plan writing involving wind energy issues.

- Planners are the local government staff most likely to be engaged in wind energy issues.

- Over two-thirds of respondents reported working with small wind energy systems, compared to just under half working with large wind energy systems. Similarly, four-fifths had an interest in small wind energy systems versus just over half reporting an interest in large wind energy systems.

- Only one-third of respondents reported addressing community wind in their communities, but almost two-thirds wanted more information on this topic.

Current Practice

- Defined size thresholds for small wind energy systems varied from 10 to 100 kW, and a few communities have added a "micro" wind category for systems under 5 kW.

- Setback requirements, height limits, and noise thresholds are common ordinance provisions for small and community wind energy systems. Large wind ordinances also commonly include abandonment clauses and require environmental review.

- More respondents (50 percent) reported small wind energy projects in their communities than large projects (24 percent) or community projects (14 percent). One-third of respondents had yet to see wind energy turbines installed in their communities.

Successes and Challenges

- Respondents found that having a good ordinance in place and strong public education and outreach efforts were important in successfully implementing wind energy systems.

- Respondents' top five most commonly identified challenges were scenic and aesthetic impacts, noise impacts, height restrictions, wildlife impacts, and property value issues. Public concerns and a lack of information about wind energy systems were also mentioned.

Information Needs

- Respondents' most commonly identified issues of importance were information on small wind energy systems, managing public concerns, noise impacts, scenic and aesthetic impacts, technical information about wind power generation, and mapping of optimal wind conditions within a jurisdiction.

- Respondents feel that there is a lack of good information and helpful resources available; they requested model and sample ordinance language, case studies, and information on potential impacts of wind turbines.

SURVEY DETAILS

Here is a more detailed description of the questions we asked and the results we received.

Survey Respondents

To begin, we asked respondents to tell us about themselves, to gain a better sense of the contexts in which they are addressing wind energy issues. Of the 180 survey responses received through September 15, 2010, 135 (84 percent) identified themselves as APA members, and 91 (57 percent) were AICP-certified planners. The majority of respondents (87 percent) were public-sector planners, with 10 percent from the private sector, 2 percent from academia, 1 percent from nonprofits, and 6 percent from other sectors, including the federal government. (Respondents could select more than one response to this question.)

All states except Alabama, Arizona, Kansas, Kentucky, Montana, New Hampshire, North Dakota, and West Virginia were represented, and a few responses came from outside the United States. Just under half of the respondents (43 percent) work in both rural and urban contexts, with 27 percent working in only an urban context and 27 percent working only in a rural setting. More than half of respondents work with small or medium-sized municipalities (29 and 27 percent, respectively), 40 percent work at the county level, and around 10 percent each work in large cities (100,000+ population), regions, and the federal government. (Respondents could select more than one category.)

Attitudes Toward Wind Energy

We asked respondents to tell us how they and their communities viewed wind energy. Respondents reported largely positive attitudes toward wind energy (85 percent versus 12 percent neutral and 4 percent negative), with 41 percent indicating a "very positive" attitude. We then asked respondents about their experiences with wind energy. Just over half reported positive experiences, with only 11 percent reporting a "very positive" experience. "Neutral" responses were 28 percent, and negative responses were 17 percent. Four percent reported a "very negative" experience. (See Figure A.1.)

When asked what had influenced their experiences with wind energy, respondents cited a general interest in environmental or sustainability issues; concerns about climate change, energy independence, and sustainability were common responses. Some respondents mentioned research performed by agencies such as APA or the American Wind Energy Association, as well as their personal research on wind energy. Public interest in, demand for, or resistance to alternative energy was a factor for some respondents. Other responses included potential cost savings associated with wind energy, awareness of communities with successful wind energy projects, concerns about the objectivity of information available from wind energy proponents, and improvements in wind energy technology.

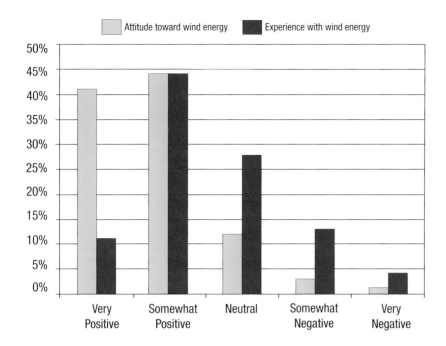

Figure A.1. *Attitude toward and experience with wind energy*

When asked to estimate levels of public interest in wind energy within their communities, respondents reported moderate (34 percent) to high (42 percent, with 18 percent reporting "very high") interest levels; one-quarter reported some to very low interest.

We then asked respondents to estimate the percentage of public opinion for and against wind energy in their community. More than half (60 percent) of responses reported more community support than opposition, 20 percent reported equal levels of support and opposition, and 15 percent reported more opposition than support. Of the responses showing a majority of community support, more than half were in the 60 to 70 percent support range.

Involvement with Wind Energy Planning

To assess the current state of wind energy planning, we asked respondents several questions about the types of wind energy planning activities with which they have been involved. Ordinance writing was the most common activity of respondents, followed by research and data collection. (See Table A.1.)

TABLE A-1. INVOLVEMENT IN WIND ENERGY PLANNING ACTIVITIES

Activity	Percent of Respondents
Ordinance writing	68
Research/data collection	56
Project review	33
Public outreach	29
Project development	19
No involvement	6

Note: Respondents were able to select more than one response.

We also asked respondents to tell us how their communities had addressed wind energy planning. The most common responses (Table A.2) were adoption or development of wind energy ordinances and codification of wind energy permitting standards. Other communities addressed wind energy in their visioning processes, established comprehensive plan policies on this topic, or conducted public outreach and education programs. Only a handful of respondents indicated that their communities were working on financial or development incentives or wind access easement issues.

TABLE A.2. HOW RESPONDENTS' COMMUNITIES HAVE ADDRESSED WIND ENERGY

Activity	Percent of Respondents
Ordinance adopted or pending	58
Permitting standards codified	29
Policies established in comprehensive plan	24
Discussed in visioning process	23
Public education or outreach programs	21
Nothing	15
Financial or development incentive programs	7
Access or easements ordinance adopted or pending	5

Note: Respondents were able to select more than one response.

Finally, we asked where wind energy expertise lay within local government staff (Table A.3), and the majority of respondents reported that planners filled this role. One-third of respondents reported that no local government staff has wind energy expertise. Consultants were used as local wind energy experts for 18 percent of respondents, with others noting the roles of engineers, building inspectors, attorneys, and energy or sustainability managers.

TABLE A.3. LOCAL GOVERNMENT STAFF WITH EXPERTISE IN WIND ENERGY

Position	Percent of Respondents
Planner	55
No local experience	33
Consultant	18
Engineer	16
Building inspector	11
Energy manager	10
Attorney	10

Note: Respondents were able to select more than one response.

Wind Energy System Scales

We asked respondents to tell us what types of wind energy systems were being implemented within their communities (Figure A.2). Most communities categorize wind energy systems based on generating capacity of a turbine or wind energy project, and we used common size thresholds of less than 100 kW for small (distributed)/residential wind energy systems, 100 kW to 1 MW for community wind energy systems, and greater than 1 MW for large or utility-scale wind energy systems. Most respondents (69 percent) reported small-scale systems in their communities; less than half (42 percent) were working with large wind systems, and just over one-third (36 percent) reported local community wind systems. For 15 percent of respondents, wind energy systems were not being addressed.

Because there is a wide range of definitions for different scales of wind energy, however, we also asked respondents to tell us if their size thresholds or definitions differed from those we provided. A number of respondents reported that their ordinances set a lower threshold for small wind energy systems; these numbers ranged from 10 to 25 kW. In addition, several respondents included a category of "micro" or "mini" wind energy systems—systems with a generating capacity of 10 kW or less. Other characteristics for differentiating among system types included turbine height and whether the energy generated by the system was used onsite or sold to other users.

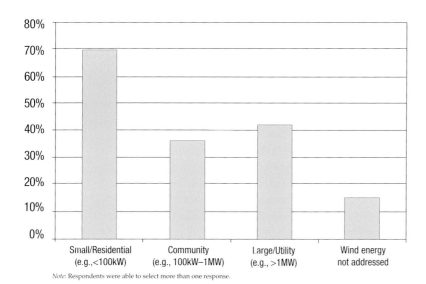

Figure A.2. Categories of wind energy systems in respondents' communities

Note: Respondents were able to select more than one response.

Wind Energy Ordinances

We also asked questions relating to the content of local ordinances regulating small, community, and large-scale wind energy systems (Table A.4). It was unclear, however, whether a lack of response meant that the community's ordinance did not include that provision or whether the community did not regulate that wind energy system scale at all.

TABLE A.4. LOCAL ORDINANCE PROVISIONS FOR WIND ENERGY

Regulation	Percent Small/ Residential	Percent Community	Percent Large/ Utility-Scale
Permitted as a primary use	11	9	11
Permitted as an accessory use	38	11	5
Permitted as a conditional or special use	35	30	31
Restricted from certain districts	24	18	20
Restricted from certain viewsheds	8	6	6
Noise thresholds	37	23	22
Setbacks	56	32	31
Height limits	51	23	21
Abandonment clause	32	16	22
Decommissioning board	9	10	15
Separate residential and nonresidential standards	18	n/a	n/a
Separate stand-alone and building-mounted system standards	19	n/a	n/a
Environmental review	n/a	n/a	22
Property value guarantee provision	n/a	n/a	1
Road maintenance or repair provision	n/a	n/a	13
Development agreement	n/a	n/a	9
Coordination with other levels of government	n/a	n/a	21
Prohibited	7	16	19
Not regulated	18	32	25
Other provisions	25	23	22

Note: Respondents were able to select more than one response.

For small wind energy systems, the five most commonly cited provisions in use by respondents' communities were setbacks (56 percent), height limits (51 percent), permitting as an accessory use (38 percent), noise thresholds (37 percent), and permitting as a conditional or special use (35 percent). Eighteen percent of respondents said small wind energy systems were not regulated in their communities.

For community wind energy systems, the five most commonly cited provisions in use by respondents' communities were setbacks (32 percent), permitting as a conditional or special use (30 percent), noise thresholds (23 percent), height limits (23 percent), and restricting from certain districts (18 percent). Thirty-two percent of respondents said community wind energy systems were not regulated in their communities.

For large wind energy systems, the five most commonly cited provisions in use by respondents' communities were permitting as a conditional or special use (31 percent), setbacks (31 percent), abandonment clauses (22 percent), environmental review (22 percent), and height limits (21 percent). One-quarter of respondents reported that large-scale wind energy systems were not regulated in their communities.

Finally, we asked about other regulations that might affect the regulation of wind energy at the local level (Table A.5).

TABLE A.5. OTHER REGULATIONS AFFECTING LOCAL REGULATION OF WIND ENERGY

Regulations	Percent of Respondents
State environmental regulations	41
State wind regulations	33
Federal environmental regulations	28
None	25
Federal military regulations	16
Neighboring jurisdiction wind regulations	12
Neighboring jurisdiction environmental regulations	4

Note: Respondents were able to select more than one response.

Though one-fourth of respondents reported that no other regulations affected local regulation of wind energy, others did indicate some level of necessary intergovernmental coordination. State environmental regulations were cited by 41 percent of respondents, while 33 percent reported state wind energy regulations in effect. Federal regulations were next on the list, with 28 percent of respondents noting federal environmental regulations and 16 percent indicating federal military regulations. Only 12 percent of respondents noted that neighboring jurisdictions' wind regulations affected their own, with 4 percent listing neighboring jurisdictions' environmental regulations. Several respondents added FAA and FCC regulations to their lists.

Wind Energy Projects

We asked respondents to list how many small, community, and large-scale wind energy projects had been developed in their communities.

Of the 146 comments submitted, half noted at least one small-scale project, while 24 percent noted large-scale projects, and only 14 percent noted community-scale projects. Among those respondents noting small-scale wind, 78 percent counted fewer than 10 projects. One-third of respondents did not have any wind energy projects currently in their communities.

Wind energy projects in respondents' communities ranged in size from small residential turbines to very large wind farms. Thirty-nine respondents reported that the largest project in the community was a residential turbine. In terms of energy output, the largest project reported was an 800 MW, 267-turbine project.

We provided a list of challenges to wind energy development, and asked respondents to tell us which they had encountered in their communities (Table A.6). The five most

TABLE A.6. CHALLENGES TO WIND ENERGY DEVELOPMENT

Challenge	Percent of Respondents
Scenic or aesthetic impacts	71
Noise impacts	63
Height restrictions	55
Wildlife impacts	48
Property values	45
Safety and security issues	36
Transmission or grid connection	36
Insufficient technical information	33
Costs	30
Lack of coordination between levels or agencies of government	26
Health issues	24
Other	22
Lack of public or private interest in wind energy development	20
Lack of financial incentives	14

Note: Respondents were able to select more than one response.

common challenges faced by respondents were scenic and aesthetic impacts (71 percent), noise impacts (63 percent), height restrictions (55 percent), wildlife impacts (48 percent), and property value issues (45 percent).

Respondents were then asked to comment about specific challenges they had faced regarding various scales of wind energy. Their comments tended to focus on these issues:

- Uncertainties in the community about wind energy

- Lack of objective information on wind energy

- Public concern over potential impacts, particularly noise, aesthetics, safety, and shadow flicker

- Prohibitive costs or lack of knowledge of financing mechanisms

- Interaction among local, state, and federal regulations

- Availability of transmission lines

We also asked respondents to comment about any barriers to wind energy implementation they had encountered within existing ordinances. This brought out these issues:

- Ordinance does not address wind turbines

- Ordinance prohibits turbines altogether or prohibits certain types of turbines (e.g., roof-mounted systems)

- Ordinance contains restrictions that limit ability to install wind turbines (e.g. height limits, minimum lot sizes, setbacks)

- Ordinance contains one set of standards for all system sizes

Wind Energy Planning Success Stories

We asked respondents to tell us about the policies, strategies, or actions regarding wind energy implementation that had worked well. Many respondents suggested having a good local ordinance that addresses wind energy, whether this meant writing a new set of regulations for wind or updating an existing wind energy ordinance. Conveying accurate and meaningful information about wind projects was also a key element of implementation. Respondents reported that taking field trips to existing wind energy projects, inviting wind energy developers to give presentations, disseminating wind energy fact sheets, and demonstrating the financial benefits of wind turbines have aided them in implementing wind energy. Finally, keeping the public involved and fostering a cooperative relationship among residents, local government, and wind energy developers were helpful.

Important Issues and Resources Needed

To assess which issues planners think are the most pressing in the field, we asked respondents to rank each item on a list of wind energy–related issues as very important, somewhat important, neutral, somewhat unimportant, or very unimportant (Table A.7). When "very important" and "somewhat important" responses were added together, the top issues were information on small wind energy systems (82 percent), managing public concerns (82 percent), noise impacts (77 percent), scenic and aesthetic impacts (75 percent), and technical information about wind power generation and mapping of optimal wind conditions within a jurisdiction (both at 73 percent). Information on large wind energy systems was cited by 53 percent of respondents, and 61 percent cited interest in information on community wind energy systems. Model ordinance language was chosen by 62 percent of respondents.

We also invited respondents to tell us what they believed were the key issues facing planners in regulating wind energy. Survey takers most commonly identified the following issues:

• A lack of accurate, nonbiased information about wind energy

• The need to educate the public about wind energy

TABLE A.7. IMPORTANT WIND ENERGY ISSUES

	Percent Very Important	Percent Somewhat Important	Percent Neutral	Percent Somewhat Unimportant	Percent Very Unimportant
Information on small or residential wind energy	40	42	8	6	5
Managing public concerns	40	42	12	4	1
Noise impacts	43	34	15	4	3
Scenic or aesthetic impacts	38	37	17	3	4
Data on or mapping of optimal wind conditions within a jurisdiction	37	36	19	5	2
Technical information about wind power generation	39	34	20	4	4
Information on or evaluation of different wind system	34	38	19	7	3
Environmental benefits of wind as a renewable resource	34	37	17	7	5
Impacts on property values	35	34	20	9	3
Wildlife impacts	34	31	28	5	2
Cost-benefit analysis	29	34	26	8	4
Model ordinance language	30	32	21	14	4
Information on community energy	19	42	22	12	5
Safety and security issues	29	30	30	7	4
Cost	22	31	33	9	5
Information on large or utility-scale wind energy	32	21	19	16	11
Transmission costs or impacts	24	27	31	15	3
Availability of utility programs	15	32	39	9	5
Health issues	23	23	34	15	5
Financing options	18	26	36	12	8

- Local ordinances that prohibit, limit, or inadequately address wind energy

- The need to understand and properly address impacts such as noise, shadow flicker, aesthetics, and wildlife habitats

- The need to strike the proper balance between the benefits and the potential impacts of wind energy

- Difficulties planning for and regulating wind energy in the face of continual technological innovation

Finally, we asked survey respondents to suggest resources and information that would assist them in implementing wind energy projects. Their responses indicated that there is a general lack of information available on planning for and regulating wind energy. In addition to basic information about wind energy, respondents were looking for model and sample ordinance language; case studies of communities that have implemented wind energy; siting guidelines; information on potential impacts of wind turbines; hard data, particularly on wind resources, performance of wind turbines, and wildlife impacts; and information on financing options.

See the survey questions at www.planning.org/research/wind.

Resource List

Anna Papke, Ann F. Dillemuth, AICP, and Suzanne Rynne, AICP

PLANNING AND REGULATING

American Planning Association. 2008. *Planning and Zoning for Renewable Energy*. Planning Advisory Service Essential Info Packet 18. Available at www.planning.org/pas/infopackets.

> Compilation of information on renewable energy includes articles, model ordinances, and local ordinances for wind energy

———. 2008. *Permitting and Standards for Wind Power*. Audioweb conference CD-ROM. Faculty: Erica Heller, AICP, Clarion Associates of Colorado, and Ron Stimmel, American Wind Energy Association. Available at www.planning.org/apastore/Search/Default.aspx?p=3889.

> CD-ROM includes an audio recording, program transcript, and PowerPoint presentation from the national conference session

———. 2009. *Renewable Local Energy*. Webinar CD-ROM. Faculty: Gary Feldman, City of Berkeley, California; Erica Heller, AICP, Clarion Associates of Colorado; Roger Taylor, National Renewable Energy Laboratory; Suzanne Rynne, AICP, American Planning Association (moderator). Available at www.planning.org/apastore/Search/Default.aspx?p=3936.

> Addresses how communities can encourage development of renewable energy infrastructure through plans and land use regulations

Andrews, C. J. 2008. "Energy Conversion Goes Local: Implications for Planners." *Journal of the American Planning Association* 74(2): 231–54.

> Proposes a new conceptual framework for energy planning, which will allow planners to better address a multitude of issues in this area of planning

Andriano, J. R. 2009. "The Power of Wind: Current Legal Issues in Siting for Wind Power." *Planning and Environmental Law* 61(5).

> Provides a legal perspective on many wind-related issues, including writing ordinances for wind energy, enacting moratoriums on wind development, and addressing aesthetic and environmental impacts

Best, A. 2010. "Transmission Boost." *Planning*. February. Available at www.planning.org/planning/2010/feb/transmission.htm.

> Discusses the implications that wind power and other alternative energy sources have for the nation's transmission grid

Davis, A., J. Rogers, and P. Frumhoff. 2008. "Putting Wind to Work." *Planning*. October. Available at www.planning.org/planning/2008/oct/puttingwind.htm.

> Gives a general overview of the state of wind energy in the United States

Duerksen, C., et al. 2011. "Renewable Energy: Wind (Small- and Large-Scale)." Section 8.1 in *Sustainable Community Development Code*. Beta Version 1.2. Denver: Rocky Mountain Land Use Institute. Available at http://law.du.edu/index.php/rmlui/rmlui-practice/code-framework/model-code.

> Presents strategies to remove obstacles, create incentives, and enact standards to encourage wind energy system development

Green, J., and M. Sagrillo. 2005. "Zoning for Distributed Wind Power—Breaking Down Barriers." National Renewable Energy Laboratory Conference Paper 500-38167. Golden, Colo.: National Renewable Energy Laboratory. Available at www.osti.gov/bridge/servlets/purl/15020001-sqRSMh.

> Addresses the challenges that multilevel, conflicting, or overlapping land-use regulations present for widespread implementation of distributed wind energy systems

Heller, E. 2008a. "Urban Wind Turbines." *Zoning Practice*. July.

> Overview of issues to take into account when drafting a zoning ordinance for small wind turbines in urban communities

———. 2008b. *Wind and Solar Production and the Sustainable Development Code.* Rocky Mountain Land Use Institute Sustainable Community Development Code Research Monologue Series: Energy. Denver: Rocky Mountain Land Use Institute. Available at http://law .du.edu/images/uploads/rmlui/rmlui-sustainable-SolarWind.pdf.

> Provides a summary of wind and solar energy technology and discusses challenges to and solutions for implementation

Homsy, G. 2007. "Earth, Wind, and Fire." *Planning.* August/September. Available at www .planning.org/planning/2007/aug/earthwind.htm.

> Discusses barriers to the development of alternative energy sources

Merriam, D. 2009. "Regulating Backyard Wind Turbines." *Vermont Journal of Environmental Law* 10(2): 291–311.

> Discusses ways in which local governments may need to regulate small wind energy systems; includes sample ordinance language

Mills, A., R. Wiser, and K. Porter. 2009. *The Cost of Transmission for Wind Energy: A Review of Transmission Planning Studies.* Berkeley, Calif.: Ernest Orlando Lawrence Berkeley National Laboratory. Available at http://eetd.lbl.gov/ea/emp/reports/lbnl-1471e.pdf.

> Examines 40 transmission planning studies and draws some conclusions about the costs of expanding the transmission grid to accommodate expanded wind energy development

National Wind Coordinating Committee (NWCC). 2005. *Technical Considerations in Siting Wind Developments: NWCC Research Meeting.* Washington, D.C., December 1–2. Available at www.nationalwind.org/assets/blog/FINAL_Proceedings.pdf.

> Summarizes the proceedings of a 2005 NWCC research meeting focused on wind turbine siting practices and their environmental impacts

———. 2005. *Wind Power Facility Siting Case Studies: Community Response.* Prepared by BBC Research and Consulting. Washington, D.C.: NWCC. Available at www.nationalwind .org/asset.aspx?AssetId=467.

> A compilation of case studies that examine the development process of nine wind energy projects; authors pay particular attention to the level of community acceptance received by the projects and the evolution of community perceptions during the development process

———. 2009. *State of the Art in Wind Siting: A Seminar.* Meeting proceedings, Washington, D.C., October 20–21. Available at www.nationalwind.org/asset.aspx?AssetId=479.

> Technical information on several topics in wind siting, including visual impacts, acoustic impacts, radar interference, property values, and icing

GUIDELINES AND TOOLKITS

American Wind Energy Association (AWEA). 2008a. *In the Public Interest: How and Why to Permit for Small Wind Systems. A Guide for State and Local Governments.* Available at: www .awea.org/documents/InThePublicInterest.pdf.

> Overview of issues related to permitting wind conversion systems within municipalities; discusses best practices for zoning for wind systems and includes a model wind ordinance

———. 2008b. *Policies to Promote Small Wind Turbines: A Menu for State and Local Governments.* Washington, D.C.: AWEA. Available at www.awea.org/learnabout/smallwind/ upload/Policies_to_Promote_Small_Wind_Turbines.pdf.

> Presents a series of policies for local governments that support increased use of wind turbines in their communities

———. 2008c. *Wind Energy Siting Handbook.* Washington, D.C.: AWEA. Available at www .awea.org/sitinghandbook/download_center.html.

> A guidebook for developers of large-scale wind energy projects that addresses the environmental aspects of siting

Asmus, P., et al. 2003. *Permitting Small Wind Turbines: A Handbook—Learning from the California Experience.* Washington, D.C.: AWEA. Available at www.consumerenergycenter .org/erprebate/documents/awea_permitting_small_wind.pdf.

> Provides an overview of small wind energy systems, information on the installation process, permitting small wind systems under California regulations, a model ordinance, and a resource and reference list

Costanti, M., et al. 2006. *Wind Energy Guide for County Commissioners.* Golden, Colo.: National Renewable Energy Laboratory and U.S. Department of Energy. Available at www.nrel .gov/wind/pdfs/40403.pdf.

> Covers the development and implementation of utility-scale wind projects (greater than 600 kW), with particular emphasis on information relevant to county commissioners and other county-level officials

Kansas Energy Council. 2005. *Wind Energy Siting Handbook: Guideline Options for Kansas Cities and Counties.* Available at http://kec.kansas.gov/reports/wind_siting_hand book.pdf.

> Provides a checklist of general wind energy development issues and concerns, as well as specific regulatory options and applications templates adopted by four Kansas counties.

Kubert, C., et al. 2004. *Community Wind Financing.* Chicago: Environmental Law & Policy Center. Available at www.elpc.org/documents/WindHandbook2004.pdf.

> Discusses financing mechanisms for community wind projects; authors address debt and equity sources, federal and state programs/incentives for wind energy, tax structures, and more

National Wind Coordinating Committee (NWCC). 2002. *Permitting of Wind Energy Facilities: A Handbook.* Washington, D.C.: NWCC. Available at www.nationalwind.org/assets/ publications/permitting2002.pdf.

> Discusses elements of the permitting process, with information on both the overall structure of the process and specific strategies; includes several case studies

Wiedman, J. 2010. *Community Renewables: Model Program Rules.* Latham, N.Y.: Interstate Renewable Energy Council. Available at http://irecusa.org/wp-content/uploads/2010/11/ IREC-Community-Renewables-Report-11-16-10_FINAL.pdf.

> Guidelines addressing administrative and operating issues for community-owned renewable energy projects

Windustry. 2007. "Wind Project Calculator." Available at www.windustry.org/your-wind-project/community-wind/community-wind-toolbox/chapter-3-project-planning-and-management/wi.

> An Excel pro forma template to help would-be wind energy developers calculate costs and returns of specific projects; includes links to other financial calculators

———. 2008. "Community Wind Toolbox." Available at www.windustry.org/Community WindToolbox.

> A comprehensive guide to planning and implementing community wind projects

IMPACTS OF WIND ENERGY AND WIND TURBINES

Alberts, D. J. 2006. *A Primer for Addressing Wind Turbine Noise.* Southfield, Mich.: Lawrence Technological University. Available at www.ltu.edu/cm/attach/165D79C3-DD14-41EC-8A7F-CFA2D0C272DE/AddressingWindTurbineNoise.pdf.

> Detailed examination of sound produced by wind turbines, its impact, and the various ways to regulate it

Colby, W. D., et al. 2009. *Wind Turbine Sound and Health Effects: An Expert Panel Review.* Washington, D.C.: AWEA and Canadian Wind Energy Association. Available at www .windpoweringamerica.gov/filter_detail.asp?itemid=2487.

> A technical discussion of noise impacts of wind turbines

Hoen, B., et al. 2009. *The Impact of Wind Power Projects on Residential Property Values in the United States: A Multi-Site Hedonic Analysis*. Berkeley, Calif.: Ernest Orlando Lawrence Berkeley National Laboratory, Environmental Energy Technologies Division. Available at http://eetd.lbl.gov/ea/ems/reports/lbnl-2829e.pdf.

> Study of the effects of wind turbines upon property values found no widespread property value impacts resulting from the presence of wind turbines

Rogers, A. L., J. F. Manwell, and S. Wright. 2006. *Wind Turbine Acoustic Noise*. White paper. Amherst: University of Massachusetts Renewable Energy Research Laboratory. Available at www.ceere.org/rerl/publications/whitepapers/Wind_Turbine_Acoustic_Noise_Rev2006.pdf.

> Detailed paper on the amount and nature of sound generated by wind turbines

U.S. Department of the Interior, Bureau of Land Management (BLM). 2009. "Wind Energy Development Programmatic EIS." Available at http://windeis.anl.gov/index.cfm.

> Materials related to the BLM's assessment of the environmental impacts of wind energy projects on BLM-owned lands in the western United States

U.S. Fish and Wildlife Service (USFWS) Wind Turbine Guidelines Advisory Committee. 2010. *Wind Turbine Advisory Committee Recommendations*. Arlington, Va.: USFWS Division of Habitat and Resource Conservation. Available at www.fws.gov/habitatconservation/windpower/Wind_Turbine_Guidelines_Advisory_Committee_Recommendations_Secretary.pdf.

> Updated version of the USFWS guidelines for wind energy development includes a section on policy recommendations as well as specific guidelines

MODEL WIND ENERGY ORDINANCES

Chicago Environmental Law Clinic and Baker & McKenzie. 2003. "Model Ordinance Regulating the Siting of Wind Energy Conversion Systems in Illinois." Available at www.illinoiswind.org/resources/pdf/WindOrdinace.pdf.

> Model ordinance seeks to encourage further wind energy development in Illinois by providing a common set of standards for wind energy developers, local governments, and residents

Daniels, K. 2005. *Wind Energy Model Ordinance Options*. Albany, N.Y.: New York Energy Research and Development Authority. Available at www.powernaturally.org/programs/wind/toolkit/2_windenergymodel.pdf.

> Provides sample ordinance language for writing large wind energy system ordinances

Iowa League of Cities. 2010. "Small Wind Innovation Zone Model Ordinance." Available at www.nwipdc.org/files/Small%20Wind%20Innovation%20Zone%20Model%20Ordinance.pdf.

> Model ordinance for Iowa communities promotes use of small wind energy systems for on-site electricity use

Lawton, C. 2002. "Commercial Wind Energy Facility and Wind Access Model Ordinance." Town of Barton, Washington County, Wisconsin. Available at www.wind-watch.org/documents/wp-content/uploads/wind-energy-model-ord.doc.

> Discusses commercial wind energy facility siting ordinance issues and provides ordinance framework

Maine State Planning Office. 2009. "Model Wind Energy Facility Ordinance." Available at www.maine.gov/spo/landuse/docs/ModelWindEnergyFacilityOrdinance.pdf.

> Detailed model ordinance for wind energy facilities; includes definitions for different types of wind energy facilities based on output

Massachusetts Executive Office of Energy and Environmental Affairs and Massachusetts Division of Energy Resources. n.d. "Model Amendment to a Zoning Ordinance or By-law: Small Wind Energy Systems." Available at www.mass.gov/Eoca/docs/doer/renew/model-allow-wind-small.pdf.

> Concise model small wind systems ordinance to provide guidance to cities and towns

———. 2008. "Companion Document to Model Amendment to a Zoning Ordinance or By-law: Allowing Wind Facilities by Special Permit." Available at www.mass.gov/Eoca/docs/doer/renew/allow-wind-by-permit-companion.pdf.

    Model development standards and commentary for utility-scale and on-site projects

Michigan Department of Labor and Economic Growth, Energy Office. 2008. "Sample Zoning for Wind Energy Systems." Available at http://miwind.msue.msu.edu/uploads/files/michigan_department_of_energy_growth.pdf.

    Guidelines to assist local governments in rural areas to develop siting requirements for both on-site and utility wind energy systems

Minnesota Project, The. 2006. "Model Wind Ordinance." Available at www.mnproject.org/e-windresources-tech.html.

    Model wind ordinance developed by several Minnesota counties; companion document also available

New Hampshire Office of Energy and Planning. 2008. "Small Wind Energy Systems." Available at www.nh.gov/oep/resourcelibrary/swes/index.htm.

    Model ordinance developed to help New Hampshire communities draft small wind energy system regulations that conform to state laws

New Jersey Board of Public Utilities, Office of Clean Energy. 2007. "Draft New Jersey Small Wind Energy System Ordinance." Fifth version. Available at www.njcleanenergy.com/files/file/SmallWindModelOrdinance111907.pdf.

    Model ordinance developed as a permitted use ordinance; can also be used as a conditional use ordinance

North Carolina Wind Working Group. 2008. "Model Wind Ordinance for Wind Energy Facilities in North Carolina." Available at www.ncsc.ncsu.edu/wind/wwg/publications/NC_Model_Wind_Ordinance_June_2008_FINAL.pdf.

    Model ordinance with provisions for small, medium, and large wind energy facilities

Oregon Department of Energy. 2005. *A Model Ordinance for Energy Projects*. Version 2. Available at www.oregon.gov/ENERGY/SITING/docs/ModelEnergyOrdinance.pdf?ga=t.

    Guide for Oregon cities and counties on siting wind, solar, biomass, geothermal, and cogeneration projects, electric power transmission and distribution lines, and other large power production facilities

Pennsylvania Department of Environmental Protection, Office of Energy and Technology Development. 2006. "Model Wind Ordinance for Local Governments." Available at www.portal.state.pa.us/portal/server.pt/community/wind/10408.

    Provides model ordinance language for amendments to an existing zoning code, subdivision/land development regulations for wind energy systems, and free-standing ordinances

South Dakota Public Utilities Commission. 2008. "Draft Model Ordinance for Siting of Wind Energy Systems (WES)." Available at http://puc.sd.gov/commission/twg/WindEnergyOrdinance.pdf.

    Model ordinance for large wind energy systems

Utah State Energy Office. 2009. "Utah Model Wind Ordinance." Available at http://geology.utah.gov/sep/wind/pdf/model_wind_ordinance.pdf.

    A best-practices document for cities and counties to consider when developing their own wind ordinances

University of Wisconsin Extension and Focus on Energy. 2005. "Small Wind Energy System Model Ordinance." Available at www3.uwm.edu/Dept/shwec/publications/cabinet/energy/Small%20Wind%20Energy%20System%20Model%20Ordinance.pdf.

    Small wind energy system model ordinance for local towns and counties (predates the 2009 revisions to wind facility siting law)

LOCAL WIND ENERGY ORDINANCES

Genesee/Finger Lakes Regional Planning Council. n.d. "Resources." Availalble at www
.gflrpc.org/programareas/wind/resources.htm.

> Lists local wind energy ordinances from New York State and provides additional
> wind-related ordinances

Oteri, F. 2008. *An Overview of Existing Wind Energy Ordinances.* Technical Report NREL/
TP-500-44439. Golden, Colo.: National Renewable Energy Laboratory. Available at www
.nrel.gov/docs/fy09osti/44439.pdf.

> Includes summary and analyses of ordinances from Illinois, Kansas, Michigan, Min-
> nesota, New York, Pennsylvania, South Dakota, Wisconsin, and Utah

U.S. Department of Energy, Wind Powering America. 2011. "Wind Energy Ordinances."
Available at www.windpoweringamerica.gov/policy/ordinances.asp.

> A database of over 100 local wind energy ordinances from communities across the
> country

WEBSITES

American Wind Energy Association: www.awea.org.

> Nonprofit organization that promotes and advocates for wind energy development
> across the United States; website has extensive resources on various aspects of wind
> energy

Database of State Incentives for Renewables and Efficiency (DSIRE): www.dsireusa.org.

> Extensive database of local, state, and federal incentives and policies for renewable
> energy, organized by state

Interstate Renewable Energy Council. "Small Wind Energy." Available at www.irecusa
.org/irec-programs/small-wind-energy.

> A clearinghouse for small wind news

National Renewable Energy Laboratory. "Wind Research." Available at www.nrel.gov/wind.

> NREL specializes in research and development related to alternative energy and
> energy efficiency; this page provides links to NREL's work on wind energy

National Wind Coordinating Collaborative: www.nationalwind.org.

> Website contains many resources and reports promoting the U.S. market for com-
> mercial wind energy, with particular emphasis on transmission, wildlife, and siting
> concerns

Salkin, Patty. *Law of the Land: A Blog on Land Use and Zoning.* Archive for the "Wind
Development" Category. Available at http://lawoftheland.wordpress.com/category/
wind-development.

> Commentary on and links to recent legal cases related to wind energy development

U.S. Department of Energy. "Wind." Available at www.energy.gov/energysources/wind.htm.

> Contains links to the DOE's wind energy programs and research efforts

———. "Wind Powering America." Available at www.windpoweringamerica.gov.

> Department of Energy initiative to substantially increase the production and use of
> wind energy in the United States; website contains information on Wind Powering
> America's varied program areas, as well as additional resources on wind energy

———. "Wind Program." Available at www1.eere.energy.gov/wind.

> Website for the DOE's Office of Energy Efficiency and Renewable Energy wind power
> program

Windustry: www.windustry.org.

> Nonprofit organization dedicated to promoting community-developed and community-
> owned wind energy projects; provides a clearinghouse of information on community
> wind

References

American Epilepsy Foundation (AEF). n.d. "Epilepsy in the News." Available at www
.epilepsyfoundation.org/resources/Epilepsy-In-The-News.cfm/photosensitivity
20060306.cfm.

American Wind Energy Association (AWEA). 2008a. *American Wind Energy Association
Siting Handbook.* 2008. Available at www.awea.org/sitinghandbook.

———. 2008b. *In the Public Interest: How and Why to Permit for Small Wind Systems: A Guide for
State and Local Governments.* Available at www.awea.org/_cs_upload/issues/3482_1.pdf.

———. 2011a. AWEA Global Market Study Fact Sheet. Washington D.C.: AWEA.

———. 2011b. *U.S. Wind Industry Annual Market Report Year Ending 2010.* Washington
D.C.: AWEA.

Arnett, E. B., W. K. Brown, W. P. Erickson, J. K. Fiedler, B. L. Hamilton, T. H. Henry, A. Jain,
G. D. Johnson, J. Kerns, and R. R. Koford. 2008. "Patterns of Bat Fatalities at Wind Energy
Facilities in North America." *Journal of Wildlife Management* 72(1): 61–78.

Arnett , E. B., M. Shirmacher, M. M. P. Huso, and J. P. Hayes. 2009. *Effectiveness of Changing
Wind Turbine Cut-In Speed to Reduce Bat Fatalities at Wind Facilities: 2008 Annual Report.
Prepared for the Bats and Wind Energy Cooperative and the Pennsylvania Game Commission.*
Austin, Tex.: Bat Conservation International. April.

"Audubon Statement on Wind Power." 2006. *Audubon Magazine.* November–December.
Available at http://policy.audubon.org/audubon-statement-wind-power.

Baerwald, E. F., J. Edworthy, M. Holder, and R. M. r. Barclay. 2009. "A Large-Scale Mitigation Experiment to Reduce Bat Fatalities at Wind Energy Facilities." *Journal of Wildlife
Management* 73(7): 1077–82.

Barone, M., and D. E. Berg. 2010. *Blade Noise Research at Sandia National Labs.* Albu-
querque, N.M.: Sandia National Laboratory. Available at http://windpower.sandia
.gov/2010BladeWorkshop/PDFs/2-2-A-1-Berg-Barone.pdf.

Bastasch, M., J. van-Dam, B. Søndergaard, and A. Rogers. 2006. "Wind Turbine Noise—
An Overview." *Canadian Acoustics* 34: 7–15.

Berg, D. E., and M. Barone. 2008. "Aerodynamic and Aeroacoustic Properties of a Flatback Air-
foil (Will It Rumble or Whisper?)." Presentation at WindPower 2008, Houston, June.

Bolinger, M. 2011. *Community Wind: Once Again Pushing the Envelope of Project Finance.*
LBNL-4103E. Berkeley, Calif: Lawrence Berkeley National Laboratory. Available at
http://eetd.lbl.gov/ea/emp/reports/lbnl-4193e.pdf.

Brenner, M., S. Cazares, M. Cornwall, F. Dyson, D. Eardley, P. Horowitz, D. Long, J. Sul-
livan, J. Vesecky, and P. Weinberger. 2008. *Wind Farms and Radar.* McLean, Va.: U.S.
Department of Homeland Security.

Chief Medical Officer of Health (CMOH) of Ontario. 2010. *The Potential Health Impact of
Wind Turbines.* Toronto: Queen's Printer for Ontario. Available at www.health.gov.on.ca/
en/public/publications/ministry_reports/wind_turbine/wind_turbine.pdf.

Colby, W. D., R. Dobie, G. Leventhall, D. M. Lipscomb, R. J. McCunney, M. T. Seilo, and
B. Søndergaard. 2009. *Wind Turbine Sound and Health Effects: An Expert Panel Review.*
Washington, D.C.: AWEA and Canadian Wind Energy Association.

Danish Energy Agency (DEA). 1999. *Wind Power in Denmark: Technologies, Policies, and Results.*
Copenhagen, Denmark. Available at http://193.88.185.141/Graphics/Publikationer/
Forsyning_UK/Wind_Power99.pdf.

Daryanian, Bahman, Donna Painter, and Patrick Brin. 2009. *Renewable Energy Development
Infrastructure Project: Regulatory and Economic Analysis.* Colorado Governor's Energy
Office, September 21. Available at http://rechargecolorado.com/images/uploads/
pdfs/redi_rwbeck%5B1%5D.pdf.

Denholm, P., M. Hand, M. Jackson, and S. Ong. 2009. *Land-Use Requirements of Modern Wind Power Plants in the United States*. August. NREL/TP-6A2-45834. Golden, Colo.: National Renewable Energy Laboratory. Available at www.nrel.gov/docs/fy09osti/45834.pdf.

Des Rosiers, F. 2002. "Power Lines, Visual Encumbrance and House Values: A Microspatial Approach to Impact Measurement." *Journal of Real Estate Research* 23(3): 275–301.

Devine-Wright, P. 2005. "Beyond NIMBYism: Towards an Integrated Framework for Understanding Public Perceptions of Wind Energy." *Wind Energy* 7: 125–39.

Edenhofer, O., R. Pichs-Madruga, Y. Sokona, K. Seyboth, D. Arvizu, T. Bruckner, J. Christensen, et al. 2011. "Summary for Policy Makers." In *IPCC Special Report on Renewable Energy Sources and Climate Change Mitigation*, ed. O. Edenhofer, R. Pichs-Madruga, Y. Sokona, K. Seyboth, P. Matschoss, S. Kadner, T. Zwickel, et al. New York: Cambridge University Press. Available at www.ipcc-wg3.de/publications/special-reports/srren.

Elliot, D., M. Schwartz, S. Haymes, D. Heimiller, G. Scott, and L. Flowers. 2010. "80 and 100 Meter Wind Energy Resource Potential for the United States." NREL/PO-550-48036. Presented at Windpower 2010 in Dallas, Texas, May 23–26. Golden, Colo.: National Renewable Energy Laboratory. Available at www.windpoweringamerica.gov/pdfs/wind_maps/poster_2010.pdf.

Encraft. 2009. Warwick Wind Trials Project, final report. Available at www.warwickwindtrials.org.uk/2.html.

Erickson, W. P., G. D. Johnson, and D. P. Young Jr. 2005. *A Summary and Comparison of Bird Mortality from Anthropogenic Causes with an Emphasis on Collisions*. PSW-GTR-191. Washington, D.C.: U.S. Forest Service.

Fingersh, L., M. Hand, and A. Laxson. 2006. *Wind Turbine Design Cost and Scaling Model*. NREL/TP-500-40566. December. Golden, Colo.: National Renewable Energy Laboratory. Available at www.nrel.gov/wind/pdfs/40566.pdf.

Fink, Sari, Kevin Porter, and Jennifer Rogers. 2010. *The Relevance of Generation Interconnection Procedures to Feed-in Tariffs in the United States*. October. Golden, Colo.: National Renewable Energy Laboratory. Available at www.nrel.gov/docs/fy11osti/48987.pdf.

Firestone, J., W. Kempton, and A. Krueger. 2009. "Public Acceptance of Offshore Wind Power Projects in the USA." *Wind Energy* 12: 183–202.

Forsythe, T., J. Gilbert, and P. Tu. 2000. *Economics of Grid Connected Small Wind Turbines in the Domestic Market*. NREL/CP-500-26975. Golden, Colo.: National Renewable Energy Laboratory. Available at www.nrel.gov/docs/fy00osti/26975.pdf.

Global Wind Energy Council (GWEC). 2011. *Global Wind Report: Annual Market Update 2010*. Brussels, Belgium: GWEC. Available at www.gwec.net/index.php?id=180.

Gross, R., P. Heptonstall, D. Anderson, T. Green, M. Leach, and J. Skea. 2006. *The Costs and Impacts of Intermittency: An Assessment of the Evidence on the Costs and Impacts of Intermittent Generation on the British Electricity Network*. London: Imperial College London.

Hoen, B., R. Wiser, P. Cappers, M. Thayer, and G. Sethi. 2009. "The Impact of Wind Power Projects on Residential Property Values in the United States: A Multi-Site Hedonic Analysis." LBNL-2829E. Berkeley, Calif.: Lawrence Berkeley National Laboratory.

Hohmeyer, O., D. Mora, and F. Wetzig. 2005. *Wind Energy: The Facts*. Volume 4. Brussels, Belgium: European Wind Energy Association.

Huber, S., and R. Horbaty. 2010. *State-of-the-Art Report, IEA Wind Task 28: Social Acceptance of Wind Energy*. Paris: International Energy Agency. Available at www.socialacceptance.ch/images/IEA_Wind_Task_28_technical_report_final_20110208.pdf.

Interstate Renewable Energy Council (IREC). 2009. *Model Interconnection Procedures: 2009 Edition*. Available at http://irecusa.org/wp-content/uploads/2010/01/IREC-Interconnection-Procedures-2010final.pdf.

Jones, C., and R. Eiser. 2009. "Identifying Predictors of Attitudes Towards Local Onshore Development with Reference to an English Case Study." *Energy Policy* 37(11): 4604–14.

Junginger M., A. Faaij, and W. C. Turkenburg. 2004. "Cost Reduction Prospects for Offshore Wind Farms." *Wind Engineering* 28: 97–118.

Kansas Energy Council. 2005. *Wind Energy Siting Handbook: Guideline Options for Kansas Cities and Counties*. Special report 2005-1. April 5. Available at http://kec.kansas.gov/reports/wind_siting_handbook.pdf.

Kanteralis, C., and J. G. Walker. 1988. "The Identification and Subjective Effect of Amplitude Modulation in Diesel Engine Exhaust Noise." *Journal of Sound and Vibration* 120(2): 297–302.

Kerry & Curlinger, LLC. n.d. "What Kills Birds?" Available at www.currykerlinger.com/birds.htm.

Krohn, S., ed., with P.-E. Morthost and S. Awerbuch. 2009. *The Economics of Wind Energy: A Report by the European Wind Energy Association*. Brussels, Belgium: European Wind Energy Association. Available at www.ewea.org/index.php?id=201.

Krug, F., and B. Lewke. 2009. "Electromagnetic Interference on Large Wind Turbines." *Energies* 2: 1118–29.

Lantz, E., and S. Tegen. 2008. *Variables Affecting Economic Development of Wind Energy*. NREL/CP-500-43506. Golden, Colo.: National Renewable Energy Laboratory. Available at www.nrel.gov/docs/fy08osti/43506.pdf.

———. 2009. *Economic Development Impacts of Community Wind Projects: A Review and Empirical Evaluation*. NREL/CP-500-45555. Golden, Colo.: National Renewable Energy Laboratory. Available at www.nrel.gov/docs/fy09osti/45555.pdf.

Lutz, T., A. Herrig, W. Wörz, M. Kamruzzaman, and E. Krämer. 2006. "Design and Wind-Tunnel Verification of Low-Noise Airfoils for Wind Turbines." *AIAA Journal* 45: 779–85.

Massachusetts, State of, Department of Energy Resources and Executive Office of Environmental Affairs. 2008. Massachuetts Model Amendment to a Zoning Ordinance or Bylaw. October. Available at www.mass.gov/Eoeea/docs/doer/renewables/wind/model-wind-bylaw-0810.pdf and www.mass.gov/Eoca/docs/doer/renew/model-allow-wind-by-permit.pdf.

Matthews, J., J. Pinto, and C. Sarno. 2007. "Stealth Solutions to Solve the Radar-Wind Farm Interaction Problem." Presentation at the Loughborough Antennas and Propagation conference, Loughborough, U.K., April 2–7. Available at http://ieeexplore.ieee.org/stamp/stamp.jsp?tp=&arnumber=4218476.

McCunney, R. J., and J. Meyer. 2007. "Occupational Exposure to Noise." In *Environmental and Occupational Medicine*. 4th ed. Baltimore: Lippincott Williams and Wilkins.

McElfish, James M., Jr., and Sara Gersen. 2011. *State Enabling Legislation for Commercial-Scale Wind Power Siting and the Local Government Role*. Washington, D.C.: Environmental Law Institute.

McLaren Loring, J. 2007. "Wind Energy Planning in England, Wales and Denmark: Factors Influencing Project Success." *Energy Policy* 35(4): 2648–60.

Meyer, David, and Richard Sedano. 2002. "Transmission Siting and Permitting." In *National Transmission Grid Study Issue Papers*. Washington, D.C.: U.S. Department of Energy. Available at http://certs.lbl.gov/ntgs/issuepapers_print.pdf.

National Health and Medical Research Council (NHMRC) of the Australian Government. 2010. *Wind Turbines and Health: A Rapid Review of Evidence*. Canberra: NHMRC. Available at www.nhmrc.gov.au/_files_nhmrc/publications/attachments/new0048_evidence_review_wind_turbines_and_health.pdf.

National Research Council (NRC). 2007. *Environmental Impacts of Wind-Energy Projects*. Washington, D.C.: National Academies Press.

National Wind Coordinating Committee (NWCC). 2005. *Wind Energy Facility Siting Case Studies: Community Response*. June. Available at www.nationalwind.org/assets/publications/NWCC_Siting_Case_Studies_Final.pdf.

———. 2006. "State Siting and Permitting of Wind Energy Facilities." April. Available at www.nationalwind.org/assets/publications/Siting_Factsheets.pdf.

———. 2009. "State of the Art in Wind Siting: A Seminar." October 20–21. Available at www. nationalwind.org/assets/publications/NWCC_Siting_Case_Studies_Final.pdf.

———. 2010. "Wind Turbine Interactions with Birds, Bats, and Their Habitats: A Summary of Research Results and Priority Questions." Available at https://www.nationalwind .org/assets/publications/Birds_and_Bats_Fact_Sheet_.pdf.

Noble Environmental Power. n.d. "Wind Fact Sheet no. 4: Shadow Flicker." Available at www .noblepower.com/faqs/documents/06-08-23NEP-ShadowFlicker-FS4-G.pdf.

Pasqualetti, M. J. 2002. "Living with Wind Power in a Hostile Landscape." Pp. 153–72 in *Wind Power in View: Energy Landscapes in a Crowded World*, ed. M. Pasqualetti, P. Gipe, and R. W. Righter. San Diego: Academic Press.

Pedersen, E., F. Van den berg, R. Bakker, and J. Bouma. 2009. "Response to Noise from Modern Wind Farms in the Netherlands." *Journal of the Acoustical Society of America* 126(2): 634–43.

Pedersen, E., and K. P. Waye. 2007. "Wind Turbine Noise, Annoyance and Self-Reported Health and Well-Being in Different Living Environments." *Occupational and Environmental Medicine* 64: 480–86.

Pierpont, N. 2010. *Wind Turbine Syndrome: A Report on a Natural Experiment*. Santa Fe, N.M.: K-Selected Books.

Reategui, S., and S. Tegen. 2008. *Economic Development Impacts of Colorado's First 1000 Megawatts of Wind Energy.* NREL/CP-500-43505. Golden, Colo.: National Renewable Energy Laboratory. Available at www.nrel.gov/docs/fy08osti/43505.pdf.

Rhode Island, State of. 2009. *Terrestrial Wind Turbine Siting Report.* January 13. Available at www.dem.ri.gov/cleannrg/pdf/terrwind.pdf.

Schwartz, M., D. Heimiller, S. Haymes, and W. Musial. 2010. *Assessment of Offshore Wind Energy Resources for the United States.* NREL/TP-500-45889. June. Golden, Colo.: National Renewable Energy Laboratory. Available at www.nrel.gov/docs/fy10osti/45889.pdf.

Shaffer, J. A., and D. H. Johnson. 2008. "Displacement Effects of Wind Developments on Grassland Birds in the Northern Great Plains." Presentation at Wind Wildlife Research Meeting VII, Milwaukee, October 28–29.

Short, L. 2002. "Wind Power and English Landscape Identity." Pp. 43–58 in *Wind Power in View: Energy Landscapes in a Crowded World*, ed. M. Pasqualetti, P. Gipe, and R. W. Righter. San Diego: Academic Press.

Simon, A. M. 1996. *A Summary of Research Conducted into Attitudes to Wind Power from 1990–1996.* London: Planning and Research for the British Wind Energy Association.

Simons, R. 2006. *Peer Reviewed Evidence on Property Value Impacts by Source of Contamination: When Bad Things Happen To Good Property.* Washington, D.C.: Environmental Law Institute Press. May.

Sims, S., and P. Dent. 2007. "Property Stigma: Wind Farms Are Just the Latest Fashion." *Journal of Property Investment and Finance* 25(6): 626–51.

Sims, S., P. Dent, and G. R. Oskrochi. 2008. "Modeling the Impact of Wind Farms on House Prices in the UK." *International Journal of Strategic Property Management* 12(4): 251–69.

Smith, J. C., M. R. Milligan, E. A. DeMeo, and B. Parsons. 2007. "Utility Wind Integration and Operating Impact State of the Art." *IEEE Transactions on Power Systems* 22(3): August 2007: 900–908.

Swofford, J., and M. Slattery. 2010. "Public Attitudes of Wind Energy in Texas: Local Communities in Close Proximity to Wind Farms and Their Effect on Decision-making." *Energy Policy* 38: 2508–19.

U.K. Energy Research Centre (UKERC). 2010. *Great Expectations: The Cost of Offshore Wind in UK Waters—Understanding the Past and Projecting the Future.* September. London: UKERC. Available at www.ukerc.ac.uk/support/tiki-download_file.php?fileId=1164.

U.S. Department of Energy (DOE). 2008. *20% Wind Energy by 2030: Increasing Wind Energy's Contribution to U.S. Electricity Supply*. DOE/GO-102008-2567. Washington, D.C.: DOE. Available at www.nrel.gov/docs/fy08osti/41869.pdf.

U.S. Energy Information Administration (EIA). 2010. "International Energy Statistics: Electricity Generation Statistics." Available at www.eia.doe.gov/cfapps/ipdbproject IEDIndex3.cfm?tid=2&pid=2&aid=12.

Van den Berg, G. 2004. "Effects of the Wind Profile at Night on Wind Turbine Sound." *Journal of Sound and Vibration* 277: 955–70.

———. 2008. "Wind Turbine Power and Sound in Relation to Atmospheric Stability." *Wind Energy* 11(2): 151–69.

Van der Horst, D. 2007. "NIMBY or Not? Exploring the Relevance of Location and the Politics of Voiced Opinions in Renewable Energy Siting Controversies." *Energy Policy* 35(5): 2705–14.

Vestas Wind Systems A/S. 2006. *Life Cycle Assessment of Electricity Delivered from an Onshore Power Plant Based on Vestas V82-1.65 MW Turbines*. Randers, Denmark: Vestas.

Voorspools, K. 2000. "Energy Content and Indirect Greenhouse Gas Emissions Embedded in 'Emission-free' Power Plants: Results for the Low Countries." *Applied Energy* 67(3): 307–30.

Warren, C., C. Lumsden, S. O'Dowd, and R. Birnie. 2005. "'Green on Green': Public Perceptions of Wind Power in Scotland and Ireland." *Journal of Environmental Planning and Management* 48(6): 853–75.

Windustry. n.d.a. *Community Wind Toolbox*. Chapter 5: Siting Guidelines and Chapter 7: Leases and Easements. Available at http://windustry.advantagelabs.com/sites/windustry.org/files/Siting.pdf and www.windustry.org/your-wind-project/community-wind/community-wind-toolbox/chapter-7-leases-and-easements/community-win.

Windustry. n.d.b. *Wind Energy Easements and Leases: Bibliography*. Available at www.windustry.org/sites/windustry.org/files/LandEBibliography.pdf. See also www.windustry.org/leases.

Wiser, R., and M. Bolinger. 2010. *2009 Wind Technologies Market Report*. DOE/GO-102010-3107. August. Washington, D.C.: U.S. Department of Energy, Office of Energy Efficiency and Renewable Energy. Available at www.nrel.gov/docs/fy10osti/48666.pdf.

———. 2011. *2010 Wind Technologies Market Report*. DOE/GO-102011-3322. Washington, D.C.: U.S. Department of Energy, Office of Energy Efficiency and Renewable Energy. Available at www1.eere.energy.gov/windandhydro/pdfs/51783.pdf.

Wiser, R., M. Bolinger, and G. Barbose. 2007. "Using the Federal Production Tax Credit to Build a Durable Market for Wind Power in the United States." LBNL-63583. Berkeley, Calif.: Lawrence Berkeley National Laboratory. Available at http://eetd.lbl.gov/ea/emp/reports/63583.pdf.

Wolsink, M. 2006. "Invalid Theory Impedes Our Understanding: A Critique on the Persistence of the Language of NIMBY." *Transactions of the Institute of British Geographers* 31: 85–91.

———. 2007. "Planning of Renewables Schemes: Deliberative and Fair Decision-Making on Landscape Issues instead of Reproachful Accusations of Non-cooperation." *Energy Policy* 35(5): 2962.

World Wind Energy Association (WWEA). 2010. *World Wind Energy Report*. Available at www.wwindea.org/home/images/stories/pdfs/worldwindenergyreport2010_s.pdf.

Wüstenhagen, R., M. Wolsink, and M. Bürer. 2007. "Social Acceptance of Renewable Energy Innovation: An Introduction to the Concept." *Energy Policy* 35(5): 2683–91.

Zoellner, J., P. Schweizer-Ries, and C. Wemheuer. 2008. "Public Acceptance of Renewable Energies: Results from Case Studies in Germany." *Energy Policy* 36(11): 4136–41.

APA American Planning Association

Making Great Communities Happen

The American Planning Association provides leadership in the development of vital communities by advocating excellence in community planning, promoting education and citizen empowerment, and providing the tools and support necessary to effect positive change.

523/524. Crossroads, Hamlet, Village, Town (revised edition). Randall Arendt. April 2004. 142pp.

525. E-Government. Jennifer Evans-Cowley and Maria Manta Conroy. May 2004. 41pp.

526. Codifying New Urbanism. Congress for the New Urbanism. May 2004. 97pp.

527. Street Graphics and the Law. Daniel Mandelker with Andrew Bertucci and William Ewald. August 2004. 133pp.

528. Too Big, Boring, or Ugly: Planning and Design Tools to Combat Monotony, the Too-big House, and Teardowns. Lane Kendig. December 2004. 103pp.

529/530. Planning for Wildfires. James Schwab and Stuart Meck. February 2005. 126pp.

531. Planning for the Unexpected: Land-Use Development and Risk. Laurie Johnson, Laura Dwelley Samant, and Suzanne Frew. February 2005. 59pp.

532. Parking Cash Out. Donald C. Shoup. March 2005. 119pp.

533/534. Landslide Hazards and Planning. James C. Schwab, Paula L. Gori, and Sanjay Jeer, Project Editors. September 2005. 209pp.

535. The Four Supreme Court Land-Use Decisions of 2005: Separating Fact from Fiction. August 2005. 193pp.

536. Placemaking on a Budget: Improving Small Towns, Neighborhoods, and Downtowns Without Spending a Lot of Money. Al Zelinka and Susan Jackson Harden. December 2005. 133pp.

537. Meeting the Big Box Challenge: Planning, Design, and Regulatory Strategies. Jennifer Evans-Cowley. March 2006. 69pp.

538. Project Rating/Recognition Programs for Supporting Smart Growth Forms of Development. Douglas R. Porter and Matthew R. Cuddy. May 2006. 51pp.

539/540. Integrating Planning and Public Health: Tools and Strategies To Create Healthy Places. Marya Morris, General Editor. August 2006. 144pp.

541. An Economic Development Toolbox: Strategies and Methods. Terry Moore, Stuart Meck, and James Ebenhoh. October 2006. 80pp.

542. Planning Issues for On-site and Decentralized Wastewater Treatment. Wayne M. Feiden and Eric S. Winkler. November 2006. 61pp.

543/544. Planning Active Communities. Marya Morris, General Editor. December 2006. 116pp.

545. Planned Unit Developments. Daniel R. Mandelker. March 2007. 140pp.

546/547. The Land Use/Transportation Connection. Terry Moore and Paul Thorsnes, with Bruce Appleyard. June 2007. 440pp.

548. Zoning as a Barrier to Multifamily Housing Development. Garrett Knaap, Stuart Meck, Terry Moore, and Robert Parker. July 2007. 80pp.

549/550. Fair and Healthy Land Use: Environmental Justice and Planning. Craig Anthony Arnold. October 2007. 168pp.

551. From Recreation to Re-creation: New Directions in Parks and Open Space System Planning. Megan Lewis, General Editor. January 2008. 132pp.

552. Great Places in America: Great Streets and Neighborhoods, 2007 Designees. April 2008. 84pp.

553. Planners and the Census: Census 2010, ACS, Factfinder, and Understanding Growth. Christopher Williamson. July 2008. 132pp.

554. A Planners Guide to Community and Regional Food Planning: Transforming Food Environments, Facilitating Healthy Eating. Samina Raja, Branden Born, and Jessica Kozlowski Russell. August 2008. 112pp.

555. Planning the Urban Forest: Ecology, Economy, and Community Development. James C. Schwab, General Editor. January 2009. 160pp.

556. Smart Codes: Model Land-Development Regulations. Marya Morris, General Editor. April 2009. 260pp.

557. Transportation Infrastructure: The Challenges of Rebuilding America. Marlon G. Boarnet, Editor. July 2009. 128pp.

558. Planning for a New Energy and Climate Future. Scott Shuford, Suzanne Rynne, and Jan Mueller. February 2010. 160pp.

559. Complete Streets: Best Policy and Implementation Practices. Barbara McCann and Suzanne Rynne, Editors. March 2010. 144pp.

560. Hazard Mitigation: Integrating Best Practices into Planning. James C. Schwab, Editor. May 2010. 152 pp.

561. Fiscal Impact Analysis: Methodologies for Planners. L. Carson Bise II. September 2010. 68pp.

562. Planners and Planes: Airports and Land-Use Compatibility. Susan M. Schalk, with Stephanie A. D. Ward. November 2010. 72pp.

563. Urban Agriculture: Growing Healthy, Sustainable Places. Kimberley Hodgson, Marcia Caton Campbell, and Martin Bailkey. January 2011. 148pp.

564. E-Government (revised edition). Jennifer Evans-Cowley and Joseph Kitchen. April 2011. 108pp.

565. Assessing Sustainability: A Guide for Local Governments. Wayne M. Feiden, with Elisabeth Hamin. July 2011. 108pp.

566. Planning for Wind Energy. Suzanne Rynne, Larry Flowers, Eric Lantz, and Erica Heller, Editors. November 2011. 140pp.